AIM for Success Practice Sheets

Beginning Algebra

AF885342

EIGHTH EDITION

Richard N. Aufmann
Palomar College

Joanne S. Lockwood
Nashua Community College

Prepared by

Christine S. Verity

BROOKS/COLE
CENGAGE Learning

Australia • Brazil • Japan • Korea • Mexico • Singapore • Spain • United Kingdom • United States

BROOKS/COLE
CENGAGE Learning

ISBN-13: 978-1-133-11226-6
ISBN-10: 1-133-11226-9

Brooks/Cole
20 Davis Drive
Belmont, CA 94002-3098
USA

Cengage Learning is a leading provider of customized learning solutions with office locations around the globe, including Singapore, the United Kingdom, Australia, Mexico, Brazil, and Japan. Locate your local office at: **www.cengage.com/global**

Cengage Learning products are represented in Canada by Nelson Education, Ltd.

To learn more about Brooks/Cole, visit **www.cengage.com/brookscole**

Purchase any of our products at your local college store or at our preferred online store **www.cengagebrain.com**

Printed in the United States of America
1 2 3 4 5 6 7 15 14 13 12 11

Contents

Name Score

Place the correct symbol, < or >, between the values of the two numbers.

1. –1 –3 **2.** –8 –5 **3.** 7 0

4. –24 42 **5.** 53 –64 **6.** –55 53

1. ______
2. ______
3. ______
4. ______
5. ______
6. ______

Answer True or False.

7. –5 > –2 **8.** 3 < –5 **9.** –12 < 14

10. –34 > 47 **11.** 19 < –31 **12.** –10 > –7

13. 53 < –71 **14.** –53 < –81 **15.** 84 > 73

7. ______
8. ______
9. ______
10. ______
11. ______
12. ______
13. ______
14. ______
15. ______

Use the roster method to write the set.

16. The natural numbers less than 4

17. The negative integers greater than or equal to –3

18. The positive integers less than 6

19. Given A = {–6, –3, 1, 4}, which elements of set A are less than –1?

20. Given B = {–9, –5, –1, 3}, which elements of set B are greater than –6?

21. Given C = {–20, 0, 1, 18}, which elements of set C are less than 1?

22. Given D = {–16, –5, 2, 3, 7, 18}, which elements of set D are greater than or equal to 2?

23. Given E = {–26, –19, –7, –5, 2, 6, 9}, which elements of set E are less than or equal to 6?

24. Given F = {–35, –15, –1, 0, 3, 6}, which elements of set F are greater than –1?

16. ______
17. ______
18. ______
19. ______
20. ______
21. ______
22. ______
23. ______
24. ______

Name Score

Find the additive inverse.

1. 7
2. –4
3. –5
4. 34
5. –28
6. 66

1. ________
2. ________
3. ________
4. ________
5. ________
6. ________

Evaluate.

7. (-3)
8. $-(7)$
9. $-(-5)$
10. $-(-13)$
11. $-|-4|$
12. $|15|$
13. $|-16|$
14. $-|19|$
15. $-|-19|$

7. ________
8. ________
9. ________
10. ________
11. ________
12. ________
13. ________
14. ________
15. ________

Place the correct symbol, < or >, between the values of the two numbers.

16. $-|21| \quad |-26|$
17. $-|22| \quad -|31|$
18. $-|14| \quad -|-35|$
19. $-|-33| \quad |54|$
20. $|40| \quad |-54|$
21. $-|28| \quad |9|$
22. $|6| \quad -|43|$
23. $-|-25| \quad |18|$
24. $|-15| \quad |4|$

16. ________
17. ________
18. ________
19. ________
20. ________
21. ________
22. ________
23. ________
24. ________

Name Score

Add.

1. $2 + (-4)$

2. $-5 + 3$

3. $-3 + (-7)$

4. $-16 + 23$

5. $9 + (-7)$

6. $-10 + (-8)$

7. $-1 + (-6)$

8. $-7 + (-7)$

9. $5 + (-10)$

10. $-9 + 3$

11. $-7 + 11$

12. $-13 + 9$

13. $6 + (-4) + (-7)$

14. $-4 + (-10) + (-11)$

15. $11 + (-3) + (-13)$

16. $-15 + (-5) + 30$

17. $10 + 54 + (-28)$

18. $-4 + (-9) + 9$

19. $-36 + (-40) + (-16)$

20. $11 + (-20) + (-30)$

21. $-11 + (-6) + 8$

22. $16 + (-18) + 5 + (-4)$

23. $-12 + (-4) + 6 + (-7)$

24. $-23 + 9 + 3 + (-16)$

25. $6 + 8 + (-13) + (-4)$

26. $-12 + (15) + (-16) + (-10)$

27. $-21 + (-33) + 27 + 18$

1. ____________
2. ____________
3. ____________
4. ____________
5. ____________
6. ____________
7. ____________
8. ____________
9. ____________
10. ____________
11. ____________
12. ____________
13. ____________
14. ____________
15. ____________
16. ____________
17. ____________
18. ____________
19. ____________
20. ____________
21. ____________
22. ____________
23. ____________
24. ____________
25. ____________
26. ____________
27. ____________

Name Score

Subtract.

1. $13 - 7$

2. $15 - 5$

3. $5 - 15$

4. $-6 - 3$

5. $-7 - 5$

6. $6 - (-4)$

7. $2 - (-2)$

8. $-9 - (-1)$

9. $-3 - (-7)$

10. $-17 - 11$

11. $-5 - 2 - 1$

12. $-4 - 5 - 11$

13. $4 - 5 - 6$

14. $10 - (-9) - 7$

15. $-3 - 10 - (-6)$

16. $12 - 9 - 11$

17. $-4 - (-5) - (-7)$

18. $6 - 7 - (-2)$

19. $-28 - (-52) - 24 - 9$

20. $44 - (-80) - 60 - 10$

21. $-17 - 36 - 62 - 14$

22. $38 - (-21) - 62 - (-17)$

23. $-36 - (-58) - 10 - 17$

24. $-16 - 37 - (-64) - 35$

25. $-19 - (-31) - (-76) - 24$

26. $21 - 33 - 57 - 14$

27. $38 - (-22) - 63 - (-9)$

1. ____________
2. ____________
3. ____________
4. ____________
5. ____________
6. ____________
7. ____________
8. ____________
9. ____________
10. ____________
11. ____________
12. ____________
13. ____________
14. ____________
15. ____________
16. ____________
17. ____________
18. ____________
19. ____________
20. ____________
21. ____________
22. ____________
23. ____________
24. ____________
25. ____________
26. ____________
27. ____________

Name Score

Multiply.

1. 11×4
2. -2×4
3. $3(-4)$
4. $-7(-2)$
5. $(8)(-2)$
6. $4(-6)$
7. $(-6)(7)$
8. $(-4)(5)$
9. $(-4)(-7)$
10. $(-2)(-9)$
11. $(0)(-6)$
12. -21×3
13. $18(-5)$
14. $3(-19)$
15. $-4(-13)$
16. $-4(27)$
17. $9(-18)$
18. $-6(-32)$
19. $-5(36)$
20. $3(-7)(2)$
21. $(-3)(5)(8)$
22. $6 \times 6 \times (-3)$
23. $(-8)(3)(-2)$
24. $-7(-6)(-5)$
25. $-4(7)(-2)$
26. $(-5)(4)(9)$
27. $4(-3)(4)$

1. ________
2. ________
3. ________
4. ________
5. ________
6. ________
7. ________
8. ________
9. ________
10. ________
11. ________
12. ________
13. ________
14. ________
15. ________
16. ________
17. ________
18. ________
19. ________
20. ________
21. ________
22. ________
23. ________
24. ________
25. ________
26. ________
27. ________

Name Score

Divide.

1. $18 \div (-6)$
2. $(-36) \div (6)$
3. $(-32) \div (-4)$
4. $7 \div (-7)$
5. $-42 \div 6$
6. $40 \div (-8)$
7. $-30 \div 5$
8. $-48 \div 6$
9. $-72 \div (-8)$
10. $-42 \div (-7)$
11. $-42 \div (-6)$
12. $-50 \div 5$
13. $-55 \div (-11)$
14. $-84 \div (-7)$
15. $-60 \div (-5)$
16. $-75 \div (-5)$
17. $105 \div 7$
18. $68 \div (-4)$
19. $72 \div (-9)$
20. $(-128) \div 8$
21. $192 \div (-12)$
22. $154 \div (-22)$
23. $-182 \div (-13)$
24. $205 \div (-5)$
25. $-120 \div (-8)$
26. $135 \div 15$
27. $168 \div (-12)$

1. ______
2. ______
3. ______
4. ______
5. ______
6. ______
7. ______
8. ______
9. ______
10. ______
11. ______
12. ______
13. ______
14. ______
15. ______
16. ______
17. ______
18. ______
19. ______
20. ______
21. ______
22. ______
23. ______
24. ______
25. ______
26. ______
27. ______

Name Score

Solve.

1. The highest elevation in Asia is Mt. Everest with an elevation of 8850 m. The lowest point is the Dead Sea with an elevation of –400 m. What is the difference between the highest and lowest points in Asia?

2. The average high temperature in Fargo, ND during the month of July is 89.6°F. The average low temperature in February is –17.6°F. Find the difference between the two average temperatures.

1. ______________

2. ______________

3. When a certain amount of salt is added to water, the freezing point of the salt water mixture is –4°C. The boiling point of the mixture is 103°C. What is the difference between the boiling point and the freezing point of the salt water mixture?

4. The temperature at which a rare element boils is 350°C. This element freezes at –29°C. Determine the difference between the temperature at which it boils and the temperature at which it freezes.

3. ______________

4. ______________

5. When the temperature is –10°F and the wind is blowing at 15 mph, the wind chill factor is –32°F. For the same temperature of –10°F and the wind blowing at 30 mph, the wind chill factor is –39°F. Find the difference between the wind chill factors.

6. The temperature at which Xenon boils is –107°C. Xenon melts at –112°C. Determine the difference between the temperatures at the boiling point and the melting point.

5. ______________

6. ______________

7. The daily low temperatures during one week were recorded as follows: 4°, –6°, 8°, –2°, –9°, –11°, and –5°. Find the average daily low temperature for the week.

8. The daily high temperatures during one week were recorded as follows: –5°, –8°, 6°, 8°, 0°, –6°, and –2°. Find the average daily high temperature for the week.

7. ______________

8. ______________

9. One golfer had a score of six under par (–6) while a second golfer had a score of 8 over par (+8). Find the difference between their scores.

10. A four-member golf team had a combined score of 18 under par (–18). Another team had a combined score of 2 over par (+2). Find the difference between their scores.

9. ______________

10. ______________

Name Score

Write as a decimal. Place a bar over repeating digits.

1. $\frac{3}{8}$

2. $\frac{7}{9}$

3. $\frac{1}{12}$

4. $\frac{7}{15}$

5. $\frac{5}{16}$

6. $\frac{3}{20}$

7. $\frac{9}{20}$

8. $\frac{19}{20}$

9. $\frac{3}{25}$

10. $\frac{7}{25}$

11. $\frac{11}{30}$

12. $\frac{15}{30}$

1. ____________

2. ____________

3. ____________

4. ____________

5. ____________

6. ____________

7. ____________

8. ____________

9. ____________

10. ____________

11. ____________

12. ____________

Name Score

Simplify.

1. $-\frac{3}{10}+\frac{1}{4}$

2. $-\frac{3}{5}+\frac{11}{20}$

3. $-\frac{5}{12}+\frac{11}{24}$

4. $-\frac{8}{9}+\frac{8}{15}$

5. $-\frac{7}{16}+\left(-\frac{3}{4}\right)+\frac{5}{8}$

6. $\frac{3}{8}+\left(-\frac{7}{12}\right)+\frac{1}{4}$

7. $-\frac{5}{9}+\frac{8}{15}+\frac{4}{5}$

8. $-9.037 + 5.7$

9. $3.27 + 5.816$

10. $\frac{1}{6}+\frac{2}{9}+\frac{5}{18}$

11. $-\frac{7}{8}+\left(-\frac{7}{12}\right)$

12. $-7.63 + 32.49 + (-89.35)$

13. $\frac{3}{4}-\frac{1}{6}$

14. $\frac{5}{6}-\frac{7}{18}$

15. $-\frac{7}{24}-\frac{5}{16}$

16. $-\frac{5}{18}-\frac{5}{6}$

17. $-\frac{3}{8}-\left(-\frac{5}{6}\right)$

18. $\frac{2}{3}-\left(-\frac{3}{10}\right)$

19. $-\frac{7}{9}+\frac{3}{4}-\frac{5}{18}$

20. $\frac{1}{3}-\frac{5}{8}-\left(-\frac{3}{4}\right)$

21. $1.92 - 6.8$

22. $-7.69 - (-5.823) - 13.513$

23. $-\frac{7}{12}-\left(-\frac{11}{24}\right)$

24. $\frac{5}{9}-\left(-\frac{5}{12}\right)+\frac{5}{18}$

1. ________
2. ________
3. ________
4. ________
5. ________
6. ________
7. ________
8. ________
9. ________
10. ________
11. ________
12. ________
13. ________
14. ________
15. ________
16. ________
17. ________
18. ________
19. ________
20. ________
21. ________
22. ________
23. ________
24. ________

Name Score

Simplify.

1. $\frac{1}{3}\times\left(-\frac{3}{5}\right)$

2. $-\frac{5}{6}\times\left(-\frac{9}{20}\right)$

3. $\left(-\frac{4}{15}\right)\left(-\frac{3}{16}\right)$

4. $\left(\frac{3}{8}\right)\left(\frac{9}{10}\right)\left(-\frac{8}{15}\right)$

5. $-\frac{6}{7}\cdot\left(-\frac{15}{22}\right)\left(\frac{28}{35}\right)$

6. $\left(-\frac{3}{4}\right)\left(\frac{12}{25}\right)\left(\frac{20}{27}\right)$

7. 1.5×2.84

8. -0.6×5.4

9. $(-1.78)(-1.2)$

10. $\frac{5}{9}\div\frac{1}{3}$

11. $\frac{9}{16}\div\left(-\frac{3}{8}\right)$

12. $-\frac{7}{10}\div\frac{21}{25}$

13. $-\frac{35}{48}\div\frac{7}{20}$

14. $\frac{21}{32}\div\frac{7}{12}$

15. $-\frac{9}{22}\div\frac{6}{11}$

16. $\frac{5}{12}\div\left(-\frac{25}{36}\right)$

17. $-\frac{8}{21}\div\left(-\frac{2}{7}\right)$

18. $-\frac{8}{9}\div\left(-\frac{2}{3}\right)$

19. $-\frac{5}{12}\div 5$

20. $-0.918\div(-0.204)$

21. $\left(-\frac{3}{7}\right)\div\left(-\frac{14}{3}\right)$

22. $6.15\div(-37.8)$
Round to the nearest thousandth.

23. $-0.2354\div(-17.45)$
Round to the nearest thousandth.

24. $-\frac{2}{3}\div\frac{4}{7}-\frac{3}{4}$

1. ____________
2. ____________
3. ____________
4. ____________
5. ____________
6. ____________
7. ____________
8. ____________
9. ____________
10. ____________
11. ____________
12. ____________
13. ____________
14. ____________
15. ____________
16. ____________
17. ____________
18. ____________
19. ____________
20. ____________
21. ____________
22. ____________
23. ____________
24. ____________

Name Score

Write as a fraction and a decimal.

1. 30% 2. 44% 3. 110%

4. 140% 5. 21% 6. 15%

1. ____________
2. ____________
3. ____________
4. ____________
5. ____________
6. ____________

Write as a fraction.

7. $22\frac{2}{9}\%$ 8. $4\frac{2}{5}\%$ 9. $34\frac{1}{4}\%$

7. ____________
8. ____________
9. ____________

Write as a decimal.

10. 3.8% 11. 16.17% 12. 32.07

10. ____________
11. ____________
12. ____________

Write as a percent.

13. 1.18 14. 2.054 15. 0.009

16. $\frac{3}{5}$ 17. $1\frac{1}{3}$ 18. $3\frac{3}{4}$

19. $\frac{13}{25}$ 20. $\frac{3}{14}$ 21. $3\frac{1}{8}$

13. ____________
14. ____________
15. ____________
16. ____________
17. ____________
18. ____________
19. ____________
20. ____________
21. ____________

Name Score

Evaluate.

1. 4^2
2. 5^3
3. -5^2
4. -3^4
5. $(-2)^4$
6. $(-3)^3$
7. $(-5)^2$
8. $(-4)^2$
9. $-(-2)^3$
10. $(-3)^3$
11. $(12)^3$
12. $(-2)^2 \cdot 4^3$
13. $(-1)^3 \cdot 2^2$
14. $(-1)^2 \cdot 2^3$
15. $(-2)^4 \cdot 5^2$
16. $(-5) \cdot 4^3$
17. $2^2 \cdot 3^2 \cdot (-4)^2$
18. $(-3) \cdot 2^4$
19. $(-3) \cdot (-2)^2$
20. $(-3) \cdot (-2)^3$
21. $2^2 \cdot 3^2 \cdot (-5)$
22. $(-2)^2 \cdot 3^2 \cdot 6$
23. $(-6) \cdot 2^3 \cdot 3^2$
24. $(-2) \cdot 3^2 \cdot (-2)^2$

1. ____________
2. ____________
3. ____________
4. ____________
5. ____________
6. ____________
7. ____________
8. ____________
9. ____________
10. ____________
11. ____________
12. ____________
13. ____________
14. ____________
15. ____________
16. ____________
17. ____________
18. ____________
19. ____________
20. ____________
21. ____________
22. ____________
23. ____________
24. ____________

Name Score

Simplify by using the Order of Operation Agreement.

1. $9-3\times 2$

2. $3^2\cdot 2-4$

3. $3(5-1)-(-2)^2$

4. $15-16\div 2^2$

5. $21-15\div 3+4$

6. $10-(-2)^3-(-1)$

7. $12+18\div(-3)-1$

8. $10-2^3-(2-5)$

9. $5-3[9-(5-3)]$

10. $-3^2+2[8\div(1+3)]$

11. $20\div 5-3^2+(-2)^3$

12. $25\div(14-3^2)+(-5)$

13. $8(-9)+[3(7-3)^2]$

14. $11+\dfrac{19-3}{3^2-1}-7$

15. $32\div\dfrac{4^2}{7-5}-(-6)$

16. $48\div 4[11-(6-1)]-(-2)^2$

17. $3\cdot[15-(8-5)]\div 6$

18. $\dfrac{(-9)+(-7)}{5^2-17}\div(3-5)$

19. $-3^2+10[20\div(2-7)]$

20. $5\cdot[20-(9-4)]\div 25$

21. $6\cdot[14-(11-9)]\div 3^2$

22. $15+21\div(-7)-3$

23. $12-3^2-(2-7)$

24. $(2.54-2.09)^2\div 0.5+0.9$

1. ______________
2. ______________
3. ______________
4. ______________
5. ______________
6. ______________
7. ______________
8. ______________
9. ______________
10. ______________
11. ______________
12. ______________
13. ______________
14. ______________
15. ______________
16. ______________
17. ______________
18. ______________
19. ______________
20. ______________
21. ______________
22. ______________
23. ______________
24. ______________

Name Score

Solve.

1. Find the complement of a 52° angle.

2. Find the complement of a 79° angle.

3. Find the complement of a 9° angle.

4. Find the complement of a 90° angle.

5. Find the supplement of a 52° angle.

6. Find the supplement of a 79° angle.

7. Find the supplement of a 9° angle.

8. Find the supplement of a 111° angle.

9. Find the supplement of a 179° angle.

10. Find the supplement of a 90° angle.

11. Angle AOB is a straight line. Find $m\angle AOC$.

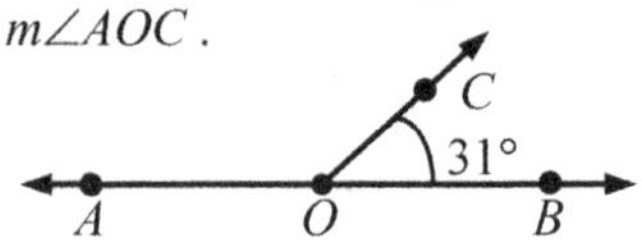

12. Angle AOB is a straight line. Find $m\angle COB$.

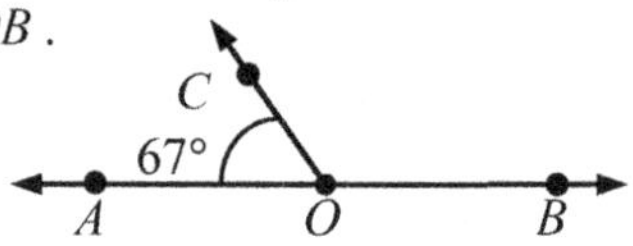

13. Find $m\angle x$.

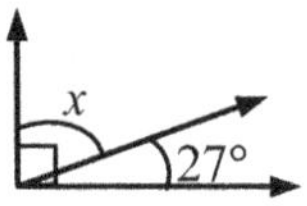

14. Find $m\angle AOB$.

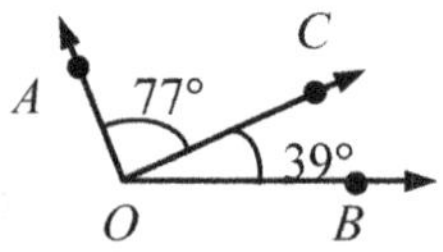

15. Find $m\angle AOC$.

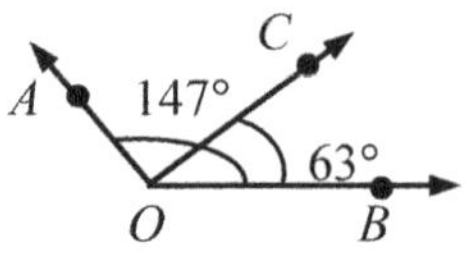

16. Find $m\angle A$.

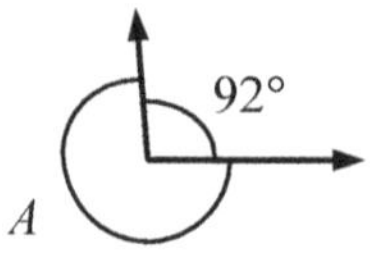

1. ________
2. ________
3. ________
4. ________
5. ________
6. ________
7. ________
8. ________
9. ________
10. ________
11. ________
12. ________
13. ________
14. ________
15. ________
16. ________

Name Score

Solve.

1. Find the perimeter of a triangle with sides 3 ft 4 in., 4 ft 7 in., and 6 ft 11 in.

2. Find the perimeter of a rectangle with length 5 ft 7 in. and width 3 ft 8 in.

3. Find the perimeter of a square with side 6 ft 2 in.

4. Find the perimeter of a square with side 37 cm.

5. Find the perimeter of a rectangle with dimensions 7 m by 11 m.

6. Find the perimeter of a triangle with sides 3.71 cm, 4.09 cm and 5.26 cm.

7. Find the circumference of a circle with radius 23 cm. Use 3.14 for π.

8. Find the circumference of a circle with diameter 1.6 m. Use 3.14 for π.

9. The wood framing for an art canvas costs $5.73 per foot. How much would the wood framing cost for a rectangular picture that measures 4 ft by 6 ft?

10. An irrigation system is installed around a circular garden that is 6 ft in diameter. If the irrigation system costs $2.59 per foot, find the cost of the system. Use 3.14 for π.

1. ____________

2. ____________

3. ____________

4. ____________

5. ____________

6. ____________

7. ____________

8. ____________

9. ____________

10. ____________

Name Score

Solve.

1. Find the area of a rectangle that measures 5 ft by 13 ft.

2. Find the area of a parallelogram with height 24.1 cm and base 13.7 cm.

3. Find the area of a parallelogram with height 9 ft and base 17 ft.

4. Find the area of a circle with radius 5 in. Use 3.14 for π.

5. Find the area of a circle with radius 7.6 m. Use 3.14 for π.

6. Find the area of a square with side 5.2 m.

7. Find the area of a square with side 2.9 yd.

8. Find the area of a triangle with height 8 cm and base 14 cm.

9. Find the area of a triangle with height 9 in. and base 19 in.

10. Find the area of a circle with diameter 4.6 m. Use 3.14 for π.

11. One side of a square room measures 15 ft. How many square yards of carpet are necessary to carpet the room. Hint: 1 yd^2 = 9 ft^2.

12. A circular stained glass window cost $42 per square foot to build. If the diameter of the window is 6 ft, find the cost to build the window. Round to the nearest dollar.

1. ______________

2. ______________

3. ______________

4. ______________

5. ______________

6. ______________

7. ______________

8. ______________

9. ______________

10. ______________

11. ______________

12. ______________

Name Score

Evaluate the variable expression when $a = -2$, $b = 3$, and $c = 4$.

1. $-2a + 3b$

2. $a^2 - 2$

3. $5(b - a)$

4. $4a + c^2$

5. $-2ab \div c$

6. $c^2 - ab$

7. $(b + c)^2$

8. $(c - a)^2 + 2b$

9. $\frac{c - a}{b}$

10. $b^2 - 2ac$

11. $-\frac{3}{4}c^2 + \frac{2}{3}b$

12. $\frac{a + 2b}{c}$

1. ______________
2. ______________
3. ______________
4. ______________
5. ______________
6. ______________
7. ______________
8. ______________
9. ______________
10. ______________
11. ______________
12. ______________

Evaluate the variable expression when $a = 3$, $b = -1$, $c = 2$ and $d = -4$.

13. $2ad - 4c$

14. $(c - b)^2 - 2d$

15. $2(b - d) + 3a$

16. $\frac{d - b}{ac}$

17. $\frac{a - d}{b}$

18. $\frac{b^2 - 4c}{-2b - a^2}$

19. $\frac{1}{4}d^2 + 3ac$

20. $\frac{5}{9}a^2 - bcd$

21. $\frac{-3cd}{ab^3}$

22. $c - \frac{1}{5}(ad - bc)$

23. $(d - a)^2 + (b - c)^2$

24. $\frac{ab - cd}{a - bc}$

13. ______________
14. ______________
15. ______________
16. ______________
17. ______________
18. ______________
19. ______________
20. ______________
21. ______________
22. ______________
23. ______________
24. ______________

Name Score

Use the given Property of the Real Numbers to complete the statement.

1. The Associative Property of Addition
$(7+3)+5=?+(3+5)$

2. The Commutative Property of Multiplication
$8\cdot 5=5\cdot ?$

3. The Multiplication Property of One
$7\cdot ?=7$

4. The Inverse Property of Addition
$x+?=0$

5. The Multiplication Property of Zero
$6\cdot 0=?$

6. The Commutative Property of Addition
$xy+wz=?+xy$

7. The Associative Property of Multiplication
$4(5x)=?\cdot x$

8. The Inverse Property of Multiplication
$\frac{1}{x}(x)=?$

9. The Addition Property of Zero
$4+?=4$

10. The Distributive Property
$3(x+2)=?\cdot x+6$

1. ____________
2. ____________
3. ____________
4. ____________
5. ____________
6. ____________
7. ____________
8. ____________
9. ____________
10. ____________

Identify the property that justifies the statement.

11. $-11+11=0$

12. $d(f+g)=df+dg$

13. $(-45y)(0)=0$

14. $a\cdot 1=a$

15. $(a+b)+c=a+(b+c)$

16. $-7+0=-7$

17. $a(bc)=(ab)c$

18. $mn\left(\frac{1}{mn}\right)=1$

19. $a+b=b+a$

20. $(a\cdot b)\cdot c=c\cdot(a\cdot b)$

11. ____________
12. ____________
13. ____________
14. ____________
15. ____________
16. ____________
17. ____________
18. ____________
19. ____________
20. ____________

Name Score

Simplify.

1. $4x+7x$

2. $10x+11x$

3. $7a-2a$

4. $-2a+9a$

5. $7ab-5ab$

6. $-10xy+15xy$

7. $9xy+(-11xy)$

8. $-13xy+9xy$

9. $-8ab+13ab$

10. $-4mn+4mn$

11. $-\frac{1}{3}x-\frac{1}{4}x$

12. $-\frac{3}{5}x+\frac{2}{10}x$

13. $\frac{1}{4}x^2-\frac{7}{12}x^2$

14. $\frac{5}{8}x-\frac{1}{2}x$

15. $-\frac{2}{3}x^2-\frac{3}{4}x^2$

16. $6a-2a+4a$

17. $-3x^2-8x^2-2x^2$

18. $-2y^2+7y^2-6y^2$

19. $6x+(-5x)+2y$

20. $7y+(-8x)+3x$

21. $7y+7x-7y$

22. $2a+(-5b)-7b+4a$

23. $9y+(-11x)-6y+4y$

24. $x^2-8x+(-3x^2)+7x$

25. $-4b+3a-9b+11a$

1. ____________
2. ____________
3. ____________
4. ____________
5. ____________
6. ____________
7. ____________
8. ____________
9. ____________
10. ____________
11. ____________
12. ____________
13. ____________
14. ____________
15. ____________
16. ____________
17. ____________
18. ____________
19. ____________
20. ____________
21. ____________
22. ____________
23. ____________
24. ____________
25. ____________

Name Score

Simplify.

1. $2(5x)$ **2.** $6(4x)$ **3.** $-3(2a)$

4. $-3(-4y)$ **5.** $-7(-8y)$ **6.** $-6(2x^2)$

7. $-7(9x^2)$ **8.** $\frac{1}{5}(5x^2)$ **9.** $\frac{1}{4}(4x^2)$

10. $\frac{1}{9}(9a)$ **11.** $\frac{1}{10}(10x)$ **12.** $-\frac{1}{3}(-3x)$

13. $-\frac{1}{5}(-5a)$ **14.** $-\frac{1}{10}(-10a)$ **15.** $(7x)\left(\frac{1}{7}\right)$

16. $(15x)\left(\frac{1}{15}\right)$ **17.** $(-8y)\left(-\frac{1}{8}\right)$ **18.** $(-20n)\left(-\frac{1}{20}\right)$

19. $\frac{1}{5}(15x)$ **20.** $\frac{1}{6}(24x)$ **21.** $-\frac{1}{3}(12x)$

22. $-\frac{1}{6}(18x)$ **23.** $-\frac{7}{8}(32a^2)$ **24.** $-\frac{3}{5}(25x^2)$

25. $-\frac{1}{3}(-15y)$ **26.** $-\frac{4}{5}(-10y)$ **27.** $(52y)\left(\frac{1}{13}\right)$

1. ____________
2. ____________
3. ____________
4. ____________
5. ____________
6. ____________
7. ____________
8. ____________
9. ____________
10. ____________
11. ____________
12. ____________
13. ____________
14. ____________
15. ____________
16. ____________
17. ____________
18. ____________
19. ____________
20. ____________
21. ____________
22. ____________
23. ____________
24. ____________
25. ____________
26. ____________
27. ____________

Name Score

Simplify.

1. $-(x+5)$ **2.** $3(2x-1)$ **3.** $4(2x-3)$

4. $-3(a+2)$ **5.** $-4(a+12)$ **6.** $-2(2y-5)$

7. $-4(3y-4)$ **8.** $3(-2x^2-12)$ **9.** $4(-5x^2-2)$

10. $-5(2y^2-3)$ **11.** $-7(3y^2-10)$ **12.** $-6(x^2+y^2)$

13. $-5(x^2-y^2)$ **14.** $-3(x^2-2y^2)$ **15.** $-2(5a^2-8b^2)$

16. $6(x^2+x-7)$ **17.** $3(x^2-2x-5)$

18. $-2(y^2+3y-5)$ **19.** $3(-a^2-a-4)$

20. $-7(-2x^2+4x+6)$ **21.** $-2(-3x^2+4x-1)$

22. $2(x^2+2xy-4y^2)$ **23.** $3(2x^2-3xy-4y^2)$

24. $-(2a^2+4a-3)$ **25.** $-(9b^2-5b+8)$

1. ______________
2. ______________
3. ______________
4. ______________
5. ______________
6. ______________
7. ______________
8. ______________
9. ______________
10. ______________
11. ______________
12. ______________
13. ______________
14. ______________
15. ______________
16. ______________
17. ______________
18. ______________
19. ______________
20. ______________
21. ______________
22. ______________
23. ______________
24. ______________
25. ______________

Name Score

Simplify.

1. $5x-3(2x+7)$
2. $3a-(4a+1)$
3. $9-(10x-4)$
4. $7-2(3x-5)$
5. $6-(8+5y)$
6. $4n-(6-3n)$
7. $3x-(10-x)$
8. $2(x+1)-4(x-6)$
9. $3(x-3)-2(x+4)$
10. $8(y-1)+2(5-2y)$
11. $5(2y-5)-2(4-y)$
12. $2(x+y)-3(x-y)$
13. $3(a+b)-(a-2b)$
14. $3[x-3(x-2)]$
15. $5[x+3(x+6)]$
16. $-3[2x+3(2-x)]$
17. $-4[x+2(6-x)]$
18. $-2[3x-(x+8)]$
19. $-5[2x-(4x-1)]$
20. $3x-2[x-3(5-x)]$
21. $-6x+4[x-6(2-x)]$
22. $-4x-3[3x-3(x+6)]-5$
23. $3a-2[3b-(2b-a)]+4b$
24. $3a-3[b-(2b-a)]+5b$
25. $3x+2(x-2y)+4(2x-5y)$
26. $4y-2(y-2x)+3(6x-y)$

1. ____________
2. ____________
3. ____________
4. ____________
5. ____________
6. ____________
7. ____________
8. ____________
9. ____________
10. ____________
11. ____________
12. ____________
13. ____________
14. ____________
15. ____________
16. ____________
17. ____________
18. ____________
19. ____________
20. ____________
21. ____________
22. ____________
23. ____________
24. ____________
25. ____________
26. ____________

Name Score

Translate into a variable expression.

1. the sum of 6 and x

2. y less 10

3. x less than 12

4. t increased by 6

5. a decreased by 1

6. 5 subtracted from y

7. z divided by 8

8. m multiplied by 11

9. 10 more than the square of x

10. 15 less than the cube of x

11. 5 times the sum of n and 8

12. the sum of two-thirds of x and 6

13. x increased by the product of 3 and x

14. the quotient of –6 and y

15. the product of –2 and t

16. the product of 2 and the sum of x and 8

17. the product of 5 and the total of y and 6

18. 10 divided by the sum of y and 5

19. 12 more than one-fourth of the square of y

20. 16 more than the product of m and –1

21. x increased by the quotient of x and 2

22. three-fifths of the product of w and 10

23. y decreased by the product of y and 2

24. the product of 7 and the total of x and 3

1. ____________
2. ____________
3. ____________
4. ____________
5. ____________
6. ____________
7. ____________
8. ____________
9. ____________
10. ____________
11. ____________
12. ____________
13. ____________
14. ____________
15. ____________
16. ____________
17. ____________
18. ____________
19. ____________
20. ____________
21. ____________
22. ____________
23. ____________
24. ____________

Name Score

Translate into a variable expression. Then simplify.

1. a number subtracted from the product of four and the number

2. a number added to the product of five and the number

3. three less than the sum of a number and twelve

4. a number increased by the difference between eleven and the number

5. a number plus the sum of the number and seven

6. a number minus the sum of the number and five

7. the sum of two-fifths of a number and three-tenths of the number

8. the difference between three-eighths of a number and one-fourth of the number

9. three less than the total of a number and nine

10. four more than the total of a number and three

11. four times the sum of five times a number and six

12. the difference between eleven times a number and seven times the number

13. twelve more than the sum of ten and a number

14. fifteen decreased by the sum of a number and five

15. three times the sum of two consecutive integers

16. nine more than the sum of two consecutive integers

17. one-half of the sum of two consecutive even integers

18. twenty more than a number added to the difference between the number and nine

19. a number subtracted from the product of the number and five

20. six times the sum of the square of a number and four

1. ____________
2. ____________
3. ____________
4. ____________
5. ____________
6. ____________
7. ____________
8. ____________
9. ____________
10. ____________
11. ____________
12. ____________
13. ____________
14. ____________
15. ____________
16. ____________
17. ____________
18. ____________
19. ____________
20. ____________

Name Score

Solve.

1. Ten gallons of paint were poured into two containers of different sizes. Use one variable to express the amount poured into each container.

2. A wire 35 feet long was cut into two pieces of different lengths. Use one variable to express the lengths of the two pieces.

3. The speed of a commercial jet is three times the speed of a corporate jet. Express the speed of the commercial jet in terms of the speed of the corporate jet.

4. A banker divided $4200 between two accounts, one paying 4% annual interest and the second paying 3.5% annual interest. Express the amount invested in the 4% account in terms of the amount invested in the 3.5% account.

5. The base of a triangle is 4 feet longer than the height of the triangle. Express the base of the triangle in terms of the height of the triangle.

6. The length of a rectangular piece of poster board is 4 times the width. Express the length of the paper in terms of the width.

7. A board 8 feet long was cut into two pieces. Express the length of the longer piece in terms of the length of the shorter piece.

8. The length of a rectangular area rug is 3 feet less than twice the width. Express the length of the rug in terms of the width of the rug.

9. The number of cherry trees in an orchard is approximately one-fifth the number of apple trees in the orchard. Express the number of cherry trees in the orchard in terms of the number of apple trees in the orchard.

10. The number of adjunct faculty at a community college is 206 more than the number of full-time faculty. Express the number of adjunct faculty in terms of the number of full-time faculty at the community college.

1. __________

2. __________

3. __________

4. __________

5. __________

6. __________

7. __________

8. __________

9. __________

10. __________

Name Score

Solve.

1. Is 2 a solution of $3a = 6$?

2. Is 5 a solution of $x - 3 = 2$?

3. Is –2 a solution of $x - 3 = -1$?

4. Is –3 a solution of $2a + 8 = 2$?

5. Is 4 a solution of $3x + 4 = 4x$?

6. Is –1 a solution of $2a + 3 = 3a + 4$?

7. Is 0 a solution of $5 - m = 3m + 3$?

8. Is 3 a solution of $3 - 2n = n + 12$?

9. Is –2 a solution of $4 - 2m = 6 - m$?

10. Is 1 a solution of $5a + 3 = 2 + 4a$?

11. Is 2 a solution of $x^2 = x + 2$?

12. Is –3 a solution of $y^2 - 1 = y + 11$?

13. Is 6 a solution of $x^2 + 3 = 9 + 5x$?

14. Is 4 a solution of $2x^2 - 1 = 4x + 15$?

15. Is $\frac{3}{4}$ a solution of $3x - 1 = 4 - x$?

16. Is $-\frac{1}{4}$ a solution of $4t + 1 = 8t + 2$?

17. Is 3 a solution of $x^2 + 4x + 4 = (x + 2)^2$?

18. Is –5 a solution of $(x - 3)^2 = x^2 + 6x + 9$?

19. Is 3 a solution of $x(x - 1) = x^2 - 3$?

20. Is $-\frac{1}{2}$ a solution of $6t + 1 = -2$?

21. Is $\frac{2}{5}$ a solution of $10m + 5 = 5m + 3$?

22. Is $\frac{1}{3}$ a solution of $2x + 1 = 2 - x$?

23. Is $\frac{1}{4}$ a solution $n + 2 = 5n + 1$?

24. Is $-\frac{1}{2}$ a solution of $2x(x - 1) = x - 1$?

1. ________
2. ________
3. ________
4. ________
5. ________
6. ________
7. ________
8. ________
9. ________
10. ________
11. ________
12. ________
13. ________
14. ________
15. ________
16. ________
17. ________
18. ________
19. ________
20. ________
21. ________
22. ________
23. ________
24. ________

Name Score

Solve and check.

1. $x+3=8$

2. $x+4=7$

3. $a-4=11$

4. $y-5=9$

5. $1+a=10$

6. $m+7=2$

7. $t+8=0$

8. $n-3=-1$

9. $x-7=-4$

10. $x-9=-5$

11. $x-5=-3$

12. $x+4=4$

13. $a-2=-7$

14. $x-4=-2$

15. $z+7=1$

16. $t-5=-3$

17. $9+a=15$

18. $-6=n+2$

19. $3=m-7$

20. $5=-10+b$

21. $-9=-3+x$

22. $12=-4+a$

23. $c+\frac{4}{5}=-\frac{1}{5}$

24. $x-\frac{1}{8}=\frac{1}{8}$

25. $x-\frac{1}{5}=\frac{2}{5}$

26. $b+\frac{1}{2}=-\frac{1}{3}$

27. $x+\frac{1}{3}=-\frac{5}{6}$

1. ____________
2. ____________
3. ____________
4. ____________
5. ____________
6. ____________
7. ____________
8. ____________
9. ____________
10. ____________
11. ____________
12. ____________
13. ____________
14. ____________
15. ____________
16. ____________
17. ____________
18. ____________
19. ____________
20. ____________
21. ____________
22. ____________
23. ____________
24. ____________
25. ____________
26. ____________
27. ____________

Name Score

Solve and check.

1. $4x = 12$
2. $5y = 30$
3. $2a = -10$
4. $3a = -15$
5. $-6m = 18$
6. $-2x = -24$
7. $-5n = -35$
8. $-48 = -8y$
9. $-42 = 6a$
10. $-18 = -6y$
11. $-\frac{y}{3} = 2$
12. $-\frac{b}{4} = 1$
13. $\frac{2}{3}y = 8$
14. $\frac{3}{5}x = 12$
15. $-\frac{3}{4}d = 9$
16. $-\frac{4}{5}m = 12$
17. $\frac{2x}{7} = 4$
18. $-\frac{2z}{5} = 8$
19. $-\frac{7z}{8} = 14$
20. $\frac{3n}{4} = 6$
21. $\frac{5x}{6} = -5$
22. $\frac{-3z}{5} = 18$
23. $-9 = -\frac{3x}{4}$
24. $-12 = \frac{-3a}{4}$
25. $\frac{2x}{3} = 1$
26. $6d - 3d = 6$
27. $9x - 5x = 20$

1. ________
2. ________
3. ________
4. ________
5. ________
6. ________
7. ________
8. ________
9. ________
10. ________
11. ________
12. ________
13. ________
14. ________
15. ________
16. ________
17. ________
18. ________
19. ________
20. ________
21. ________
22. ________
23. ________
24. ________
25. ________
26. ________
27. ________

Name Score

Solve.

1. Dan runs 3 mi. If he leaves his house at 9 AM and returns at 9:30AM, determine Dan's average rate of speed in miles per minute if he stops for 3 min to pick up trash?

2. A walk-a-thon has chosen a 20-mile course. If the walkers plan on averaging 4 mph while they are walking, and they have a 1-hour lunch break planned, how long will it take them to complete the course?

3. Judy's average walking speed is 2 kilometers per hour faster than Jean's walking speed. If Judy can walk around a 24-kilometer course in 3 h, how many hours would it take for Jean to walk the same course?

4. A large airport has a moving sidewalk that takes travelers from the middle of the airport to the end, a distance of 600 ft. If your normal walking rate is 5 ft/s and the moving sidewalk is traveling at 3 ft/s, how many seconds would it take for you to walk from the middle of the airport to the end?

5. Jason and Alexis start from the same point at the same time and cycle in opposite directions. Jason is traveling at 11 mph, and Alexis is traveling at 3 mph. After 30 min, how far apart are Jason and Alexis?

6. Christopher and Alex can paddle their canoe at a rate of 6 mph in calm water. How long will it take them to travel 12 mi against the 3 mph current of the river?

7. Skylar and Benjamin start at the same time from opposite ends of a 9-mile jogging trail and begin running toward each other. Skylar is running at the rate of 6 mph, and Benjamin is running at a rate of 4 mph. How long, in minutes, after they start will Skylar and Benjamin meet?

8. As part of the training program for the Boston Marathon, April wants to build endurance by running at a rate of 8.5 mph for 24 min. How far will April travel in that time period?

1. __________

2. __________

3. __________

4. __________

5. __________

6. __________

7. __________

8. __________

Name Score

Solve.

1. 4 is what percent of 25?

2. What percent of 75 is 15?

3. Find 16% of 30.

4. What is 30% of 90?

5. 15% of what is 60?

6. 25% of what is 8?

7. What is 12.5% of 48?

8. 20 is what percent of 50?

9. 20% of what is 8?

10. 25% of what is 16?

11. 60 is what percent of 12?

12. 24 is what percent of 6?

13. Find 18% of 40.

14. What is 27% of 14?

15. Find 30.8% of 60.

16. A company built and equipped a new suite of offices for $200,000. Of this amount, $40,000 is for equipment and furnishings. What percent of the total expenditure was for equipment and furnishings?

17. A used car dealer estimates that 65% of the company's sales will occur on Thursday, Friday or Saturday. Using this estimate, how many cars out of the 80 sold last month were sold on Thursday, Friday or Saturday?

18. Approximately 70% of the employees in a company purchase supplementary dental insurance. Using this estimate, how many of the company's 200 employees purchased dental insurance?

19. A computer dealer estimates that 12.5% of the computers sold will require service during the one year warranty period. Using this estimate, how many computers were sold in a year in which 75 new computers were serviced?

20. The number of part-time employees in a business is 40. Last year the number of part-time employees was 50. What percent decrease does this represent?

21. The number of building permits issued in a city this year was 72,000. This represents 125% of last year's number of permits. How many building permits were issued last year?

1. ____________
2. ____________
3. ____________
4. ____________
5. ____________
6. ____________
7. ____________
8. ____________
9. ____________
10. ____________
11. ____________
12. ____________
13. ____________
14. ____________
15. ____________
16. ____________
17. ____________
18. ____________
19. ____________
20. ____________
21. ____________

Name Score

Solve and check.

1. $2x+3=15$
2. $3y+5=26$
3. $4a-7=9$
4. $5m-16=19$
5. $3=4a+15$
6. $3x-7=-22$
7. $6n-9=-51$
8. $9=5+2d$
9. $17=9+4z$
10. $8-c=7$
11. $6-2w=-4$
12. $7-5x=-23$
13. $9-4t=1$
14. $16-7x=2$
15. $5y-35=0$
16. $12+3b=0$
17. $14+2m=0$
18. $-3x+7=-5$
19. $-4d+5=-31$
20. $-9x-4=-22$
21. $-11x+20=-2$
22. $-15=-12y+21$
23. $3=7-4a$
24. $2=12-5n$
25. $-33=-6b+3$
26. $-7x+2=-26$
27. $-4x-36=0$

1. ________
2. ________
3. ________
4. ________
5. ________
6. ________
7. ________
8. ________
9. ________
10. ________
11. ________
12. ________
13. ________
14. ________
15. ________
16. ________
17. ________
18. ________
19. ________
20. ________
21. ________
22. ________
23. ________
24. ________
25. ________
26. ________
27. ________

Name Score

Solve and check.

1. $7x+4=3x+32$
2. $5y+1=y+17$
3. $10m+2=9m+10$
4. $4x-3=2x+7$
5. $8a-9=2a+15$
6. $11y-3=8y-9$
7. $12b-3=4b-27$
8. $14x-1=3x-23$
9. $6a-4=a-19$
10. $4x+3=13-x$
11. $3x-4=-22-3x$
12. $5y-1=-17-3y$
13. $2b+9=5b+15$
14. $m+3=3m+11$
15. $5d-3=7d+9$
16. $5y-7=2y-7$
17. $4a+11=a+11$
18. $5-4x=7-2x$
19. $9-5n=15-2n$
20. $4+7x=12+3x$
21. $3x-5=8x$
22. $2a-15=7a$
23. $7m=4m+24$
24. $8y=4y+40$
25. $-2x-3=3x+7$
26. $-6a-1=2a+23$
27. $-7n+4=-4n-14$

1. ______
2. ______
3. ______
4. ______
5. ______
6. ______
7. ______
8. ______
9. ______
10. ______
11. ______
12. ______
13. ______
14. ______
15. ______
16. ______
17. ______
18. ______
19. ______
20. ______
21. ______
22. ______
23. ______
24. ______
25. ______
26. ______
27. ______

Name Score

Solve and check.

1. $3x + 2(x - 1) = 8$

2. $5y + 3(y - 3) = 15$

3. $8n - 5(2n + 1) = 7$

4. $10x - 3(2x + 5) = 13$

5. $8m - 5(m - 2) - 7$

6. $5a - (2a - 7) = 10$

7. $7n - 3(4n - 1) = 13$

8. $3(2x - 1) - 2 = 13$

9. $2(3b + 2) - 7 = 9$

10. $4(2 - 3y) + 3y = 26$

11. $5(1 - 2x) + 5 = 80$

12. $9x + 1 = 3(2x + 7) - 2$

13. $4y - 5 = 6 + 3(y - 1)$

14. $11x + 4 = 2(3x - 2) - 7$

15. $8 - 7x = 14 - (5x + 6)$

16. $2x - 3 = 5(3x + 2)$

17. $4n - 9 = 3(2n + 5)$

18. $4b + 2(3b - 5) = 2b + 6$

19. $x + 2(3x - 1) = 6x - 5$

20. $3 - 2a = 10 - 3(2a - 3)$

21. $8 - 4x = 11 - (5x + 6)$

22. $4b + 3(2b - 1) = 2b + 13$

23. $x + 2(3x - 2) = 6x - 2$

24. $2y - 5 = 4(y - 7) + 5$

25. $3a - 4 = 3(2a + 1) + 5$

26. $4 - (8 - 5x) = 3x - 8$

27. $6 - (4 - 7x) = 3x + 2$

1. ____________
2. ____________
3. ____________
4. ____________
5. ____________
6. ____________
7. ____________
8. ____________
9. ____________
10. ____________
11. ____________
12. ____________
13. ____________
14. ____________
15. ____________
16. ____________
17. ____________
18. ____________
19. ____________
20. ____________
21. ____________
22. ____________
23. ____________
24. ____________
25. ____________
26. ____________
27. ____________

Name Score

Solve.

Use the formula $L = P \cdot N$, where L is the loan amount, P is the monthly payment, and N is the number of months.

1. A loan of \$1152 is to be repaid in 12 equal monthly payments. Find the monthly payment.

2. The monthly payment on a \$6300 loan is \$150. Find the number of months in which the loan is repaid.

1. ____________

2. ____________

Use the equation $V = \pi r^2 h$, where V is the volume of the cylinder, r is the radius of the cylinder, and h is the height of the cylinder. Use $\frac{22}{7}$ for π.

3. The radius of a cylinder is 14 cm, and 20 cm is the height. Find the volume.

4. The volume of a cylinder is 20,790 cm^3, and the radius is 21 cm. Find the height of the cylinder.

3. ____________

4. ____________

Use the level system equation $F_1 \cdot x = F_2 \cdot (d - x)$.

5. A lever is 24 ft long. A force of 36 lb is applied to one end of the lever and a force of 12 lb is applied to the other end. Find the location of the fulcrum when the system balances.

6. A lever is 16 ft long. A force of 50 lb is applied to one end of the lever and a force of 30 lb is applied to the other. Find the location of the fulcrum when the system balances.

5. ____________

6. ____________

7. A lever is 12 ft long. At a distance of 8 ft from the fulcrum, a force of 24 lb is applied. How large a force must be applied to the other end of the lever so that the system balances?

8. A lever is 20 ft long. At a distance of 5 ft from the fulcrum, a force of 24 lb is applied. How large a force must be applied to the other end of the lever so that the system balances?

7. ____________

8. ____________

Use the equation $m_1 \cdot (T_1 - T) = m_2 \cdot (T - T_2)$ to find the final temperature of the water. m_1 is the quantity of the water at the hotter temperature, T_1 is the temperature of the hotter water, m_2 is the quantity of water at the cooler temperature, T_2 is the temperature of the cooler water, and T is the final temperature of the water after mixing.

9. A chemist mixes 600 g of water at 60°C with 200g of water at 25°C. Find the final temperature of the water after mixing.

10. A chemist mixes 400g of water at 70°C with 80g of water at 15°C. Find the final temperature of the water after mixing.

9. ____________

10. ____________

Name Score

Solve and graph the solution set.

1. $x+2>5$

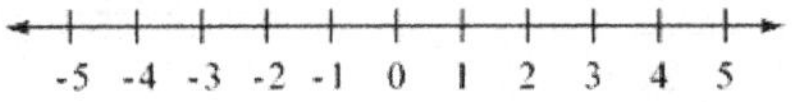

2. $y+1<4$

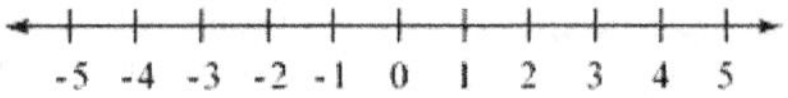

3. $x-3>-4$

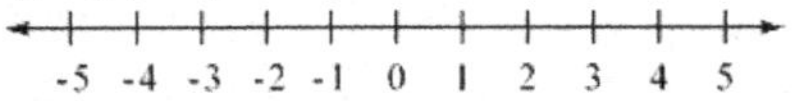

4. $x-1>-5$

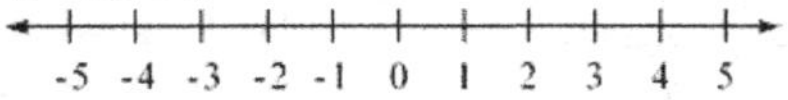

5. $n+2\geq 5$

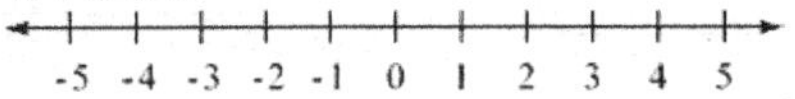

6. $x+3\geq 1$

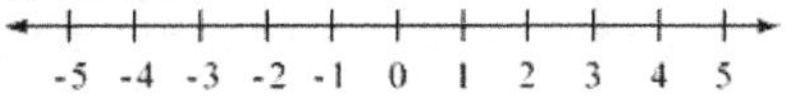

7. $x-3\leq -7$

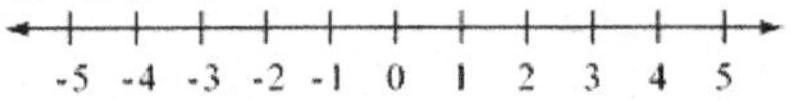

8. $y-2\leq -5$

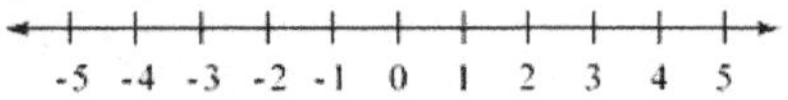

9. $4+x\geq 3$

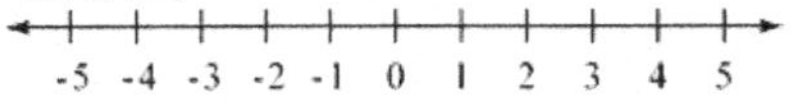

10. $-1+n\geq 0$

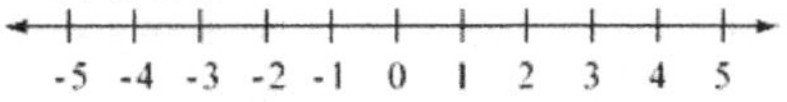

Solve.

11. $y-1\geq -11$

12. $x+6\geq -12$

13. $2x-3<x+6$

14. $7x-5\geq 6x-3$

15. $3x+2<2x-6$

16. $8x+37<x-3$

17. $3x+1\geq 2x-1$

18. $d+\frac{1}{3}<\frac{5}{6}$

19. $y-\frac{3}{8}\geq -\frac{3}{4}$

20. $3b-\frac{3}{5}\geq 2b+\frac{1}{10}$

21. $N-1.6\leq 4.8$

22. $x-2.4<1.6$

1. ________
2. ________
3. ________
4. ________
5. ________
6. ________
7. ________
8. ________
9. ________
10. ________
11. ________
12. ________
13. ________
14. ________
15. ________
16. ________
17. ________
18. ________
19. ________
20. ________
21. ________
22. ________

Name Score

Solve and graph the solution set.

1. $4y < 16$

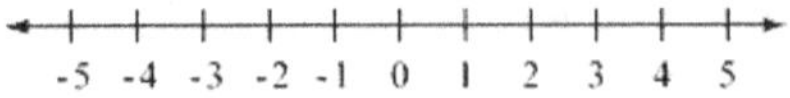

2. $3x \leq -9$

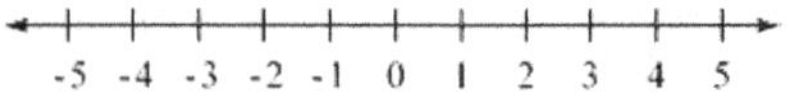

3. $6y \geq 12$

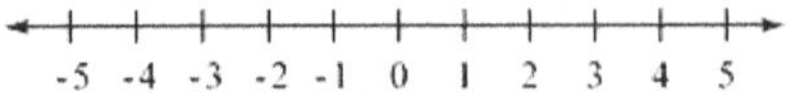

4. $15x > -30$

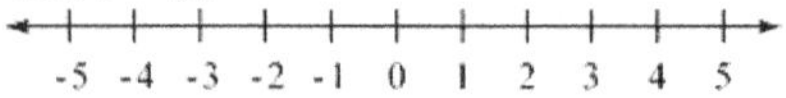

5. $8x \leq 16$

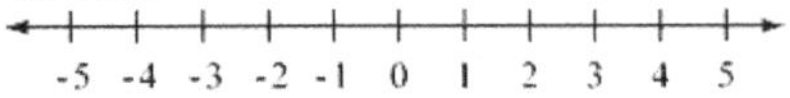

6. $2x < 0$

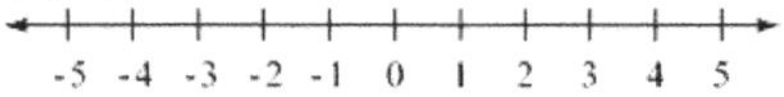

7. $-4t < 8$

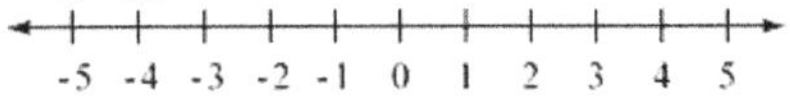

8. $-3N \leq -9$

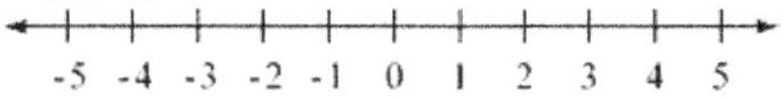

9. $-5n > 20$

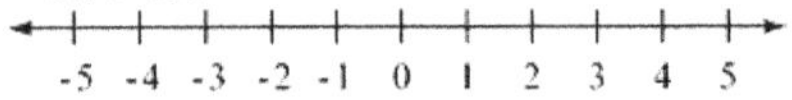

10. $-2x < 10$

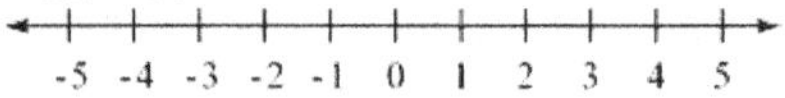

1. ________
2. ________
3. ________
4. ________
5. ________
6. ________
7. ________
8. ________
9. ________
10. ________

Solve.

11. $-4y \geq 16$

12. $2x < 8$

13. $\frac{3}{8}x \leq 6$

14. $-\frac{4}{9}b \geq -4$

15. $-\frac{4}{5}x \geq -16$

16. $\frac{2}{3}N < \frac{4}{9}$

17. $-\frac{5}{8} < -\frac{1}{4}$

18. $-\frac{3}{4}y \geq -1$

19. $-\frac{3}{11}b \geq -6$

20. $-2.5y \geq 9.2$

21. $-2.7x \geq 5.4$

22. $-2.5d > 14$

11. ________
12. ________
13. ________
14. ________
15. ________
16. ________
17. ________
18. ________
19. ________
20. ________
21. ________
22. ________

Name Score

Solve.

1. $3x - 4 < x$

2. $6x - 9 < 3x$

3. $5y - 4 > 6y$

4. $2x + 3 \geq 5x - 6$

5. $3N - 5 \geq N + 3$

6. $0.8x + 11 > 1.2x + 39$

7. $-2.5b + 3 < 1.7b + 24$

8. $3(2x - 1) > 5x + 2$

9. $4(3x + 5) > 7x - 5$

10. $2(4x - 7) > 2x + 10$

11. $5(2x + 3) > 8x + 1$

12. $2(x - 6) < 3(x - 2)$

13. $3(4x - 5) \leq 2(x - 5)$

14. $3(4 - x) > 2(2x - 1)$

15. $4(1 - x) \leq 2(3 - x)$

16. $5(3 + x) > 2(3x - 5)$

17. $4(2d - 1) > 3(1 + 3d)$

18. $3(3 - x) < 5(x + 5)$

19. $2(d - 3) + 8 \geq 3(4 - d)$

20. $4(x - 1) > 8x - 2(3x - 5)$

21. $3 - 2(2 - N) \leq 3(2 - 4N)$

22. $12 - 4(2 - x) \leq 3(x - 1)$

23. $4 - 2(4 - N) \leq 2(3 - 4N)$

24. $5 - 3(5 - 2x) \leq 2(x - 1)$

1. ______
2. ______
3. ______
4. ______
5. ______
6. ______
7. ______
8. ______
9. ______
10. ______
11. ______
12. ______
13. ______
14. ______
15. ______
16. ______
17. ______
18. ______
19. ______
20. ______
21. ______
22. ______
23. ______
24. ______

Name Score

Translate into an equation and solve.

1. The sum of a number and nine is fifteen. Find the number.

2. Two more than a number is eleven. Find the number.

3. The difference between a number and six is ten. Find the number.

4. One-half of a number is nine. Find the number.

5. Five more than four times a number is thirteen. Find the number.

6. Two times the sum of a number and three is twelve. Find the number.

7. Eight less than three times a number is equal to seven. Find the number.

8. The quotient of a number and six is the opposite of two. Find the number.

9. The sum of seven times a number and three is the opposite of eighteen. Find the number.

10. The sum of five times a number and the number is twenty-four. Find the number.

11. The difference between six times a number and three times the number is twelve. Find the number.

12. Thirty is five minus the product of five and a number. Find the number.

13. Nine is one more than the product of two and a number. Find the number.

14. The sum of two numbers is twenty-one. Twice the larger is four times the smaller number. Find the two numbers.

15. Five times a number is equal to eighteen more than twice the number. Find the number.

16. Four times the sum of a number and three equals two less than five times the number. Find the number.

1. ______
2. ______
3. ______
4. ______
5. ______
6. ______
7. ______
8. ______
9. ______
10. ______
11. ______
12. ______
13. ______
14. ______
15. ______
16. ______

Name Score

Write an equation and solve.

1. The retail selling price of a smoke detector is $36. The price is $22 more than the cost of the smoke detector. Find the cost of the smoke detector.

2. The sale price of a pair of boots is $64. This is $16 less than the original price. Find the original price.

1. ____________

2. ____________

3. Due to depreciation, the value of a motorcycle is now $2400. This is three-fourths of its original value. Find the original value.

4. An engineer's salary is $54,000. This is twice the engineer's salary twelve years ago. Find the engineer's salary twelve years ago.

3. ____________

4. ____________

5. An appliance center sold four times as many color TV sets as it sold black and white TV sets. The number of TV sets sold last year was 1200. How many color TV sets were sold?

6. The sum of the angles of a triangle is 180°. The measure of the first angle is 60°. The measure of the second angle is twice the measure of the third angle. Find the measures of the second and third angles.

5. ____________

6. ____________

7. An air conditioning repair bill was $131. This included $75 for parts and $28 for each hour of labor. Find the number of hours of labor.

8. A driver drove for 12 hours. During the first 8 hours the driver averaged 50 mph. If the driver drove 560 miles, what was the average speed during the last 4 hours of the trip?

7. ____________

8. ____________

9. A 18-foot board is cut into two pieces. Four times the length of the shorter piece is 2 ft more than the length of the longer piece. Find the length of the longer piece.

10. The engine speed of a car is 2400 rpm (revolutions per minute). This is equal to 500 rpm less than twice the drive shaft speed. Find the drive shaft speed.

9. ____________

10. ____________

Name Score

Solve.

1. The perimeter of a rectangle is 144 ft. The length of the rectangle is three times the width. Find the length and width of the rectangle.

2. The perimeter of a triangle is 42 ft. The first side is 3 ft more than the second side. The third side is 1 ft less than twice the second side. Find the measure of each side.

3. The perimeter of a rectangle is 76 m. The length of the rectangle is 5 m more than twice the width. Find the length and width of the rectangle.

4. The perimeter of a triangle is 180 ft. One side is twice the second side. The third side is 20 ft more than the second side. Find the measure of each side.

5. In an isosceles triangle, two sides are equal. The third side is 75% of the length of one of the equal sides. Find the length of each side when the perimeter is 165 cm.

6. The perimeter of a triangle is 180 cm. The length of the first side is twice the length of the second side. The length of the third side is three times the length of the second side. Find the length of each side.

7. The perimeter of a rectangle is 118 m. The width of the rectangle is 7 m less than the length. Find the length and width of the rectangle.

8. The width of a rectangle is 60% of the length. The perimeter of the rectangle is 128 ft. Find the length and width of the rectangle.

9. In an isosceles triangle, two sides are equal. The third side is 7 ft less than one of the equal sides. The perimeter is 53 ft. Find the length of each side.

10. The perimeter of a rectangle is 20 m. The length of the rectangle is 2 m more than the width. Find the measure of the width.

1. ____________

2. ____________

3. ____________

4. ____________

5. ____________

6. ____________

7. ____________

8. ____________

9. ____________

10. ____________

Name Score

Solve.

1. Find the supplement of a 111° angle.

2. Find the complement of a 90° angle.

3. Find the complement of a 9° angle.

4. Find the supplement of a 79° angle.

1. ____________

2. ____________

3. ____________

4. ____________

Find *x*.

5.

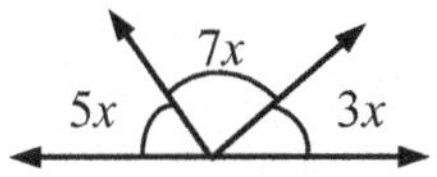

6.

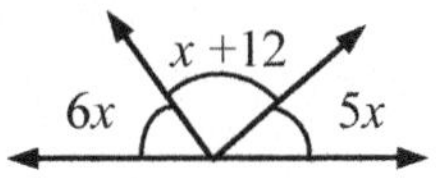

7.

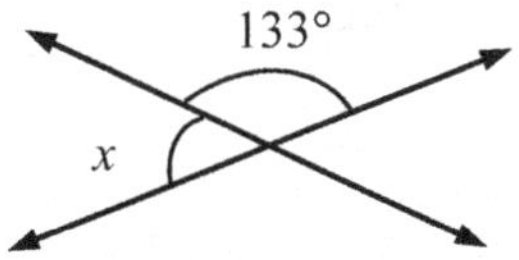

8.

5. ____________

6. ____________

7. ____________

8. ____________

Find the measures of angles *a* and *b*. *k* □ *ℓ*

9.

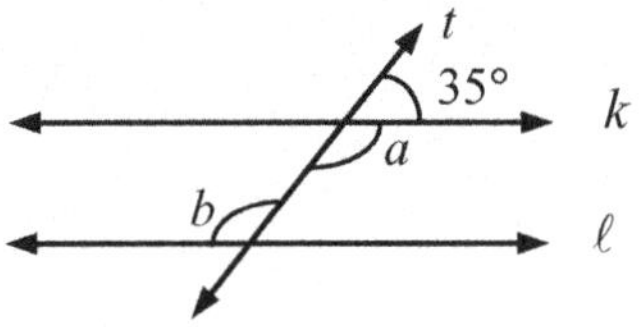

10.

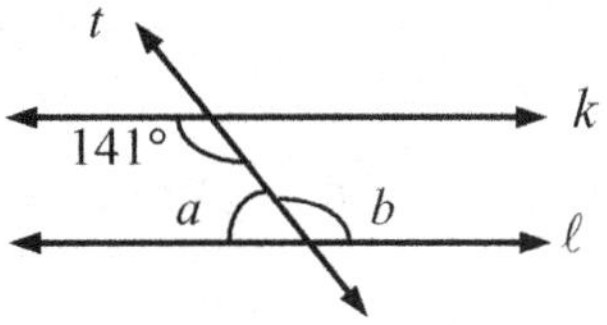

9. ____________

10. ____________

Find *x*. *k* □ *ℓ*

11.

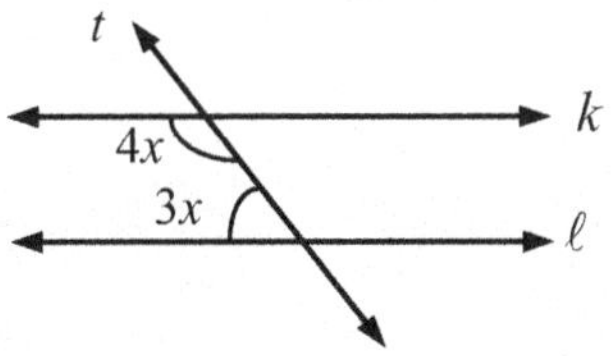

12.

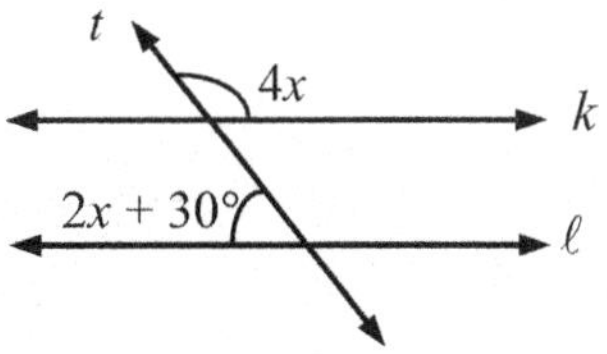

11. ____________

12. ____________

Name Score

Solve.

1. A triangle has a 40° angle and a right angle. What is the measure of the third angle?

2. A triangle has a 53° angle and a right angle. Find the measure of the third angle.

3. Two angles of a triangle measure 18° and 107°. Find the measure of the third angle.

4. Two angles of a triangle measure 68° and 43°. Find the measure of the third angle.

5. A triangle has a 121° angle and a 15° angle. What is the measure of the third angle?

6. A triangle has a 79° angle and a 72° angle. What is the measure of the third angle?

7. Two angles of a triangle measure 92° and 58°. What is the measure of the third angle?

8. A triangle has a 84° angle and a 39° angle. Find the measure of the third angle.

1. __________

2. __________

3. __________

4. __________

5. __________

6. __________

7. __________

8. __________

Name Score

Solve.

1. A grocery store buys a pound of coffee for $5.40 and uses a mark-up rate of 25%. Find the selling price.

2. A gift shop uses a mark-up rate of 48%. One lamp has a selling price of $37. Find the cost of the lamp.

3. The selling price of an action figure is $13. The mark-up rate used by the toy store is 30%. Find the cost of the action figure.

4. The manufacturer's cost for a digital camera is $180. The manufacturer then sells the digital camera for $252. What is the mark-up rate?

5. The manager of a produce stand uses a mark-up rate of 42%. What is the selling price of a bag of potatoes which cost $3?

6. A video game costs a retailer $24. Find the selling price when the mark-up rate is 35%.

7. A bakery sells a cake for $15.75. The bakery's cost to make the cake is $10.40. What is the mark-up rate?

8. The selling price for a jacket is $150. The cost to the clothing store for the jacket is $100. What is the mark-up rate?

9. The cost to a car dealer for a stereo radio is $120. The selling price of the radio is $160. What is the mark-up rate?

10. The cost to an appliance dealer for a coffee maker is $20. The dealer sells the coffee maker for $29. What is the mark-up rate?

11. The selling price for a set of golf clubs is $270. The clubs cost the golf pro $150. What is the mark-up rate?

12. The manager of an office furniture store uses a mark-up rate of 60%. What is the selling price of a file cabinet which costs $90?

1. ____________

2. ____________

3. ____________

4. ____________

5. ____________

6. ____________

7. ____________

8. ____________

9. ____________

10. ____________

11. ____________

12. ____________

Name Score

Solve.

1. A microwave oven which regularly sells for \$200 is on sale for 20% off the regular price. Find the sale price.

2. A clock radio which regularly sells for \$32 is on sale for 25% off the regular price. Find the sale price.

3. The sale price for a fire extinguisher is \$28, which is 30% off the regular price. Find the regular price.

4. The sale price for a shredder is \$56, which is 20% off the regular price. Find the regular price.

5. A bicycle which regularly sells for \$120 is on sale for \$96. Find the discount rate.

6. A cordless telephone which regularly sells for \$80 is on sale for \$60. Find the discount rate.

7. A pair of bowling shoes is on sale for \$26. This is 35% off the regular price. Find the regular price.

8. During a sale, a clothing store reduced all items 40%. What was the regular price of a blazer which was on sale for \$90?

9. A home computer which regularly sells for \$400 is on sale for \$340. Find the discount rate.

10. A bag of grass seed costs a retailer \$12. Find the selling price when the mark-up rate is 45%.

11. A tennis racket is on sale for \$24. This is $33\frac{1}{3}\%$ off the regular price. Find the regular price.

12. A lawn chair which regularly sells for \$20 is on sale for \$15.50. Find the discount rate.

1. ______________

2. ______________

3. ______________

4. ______________

5. ______________

6. ______________

7. ______________

8. ______________

9. ______________

10. ______________

11. ______________

12. ______________

Name Score

Solve.

1. A total of $8000 is deposited into two simple interest accounts. On one account, the annual interest rate is 10%, while on the second account the annual simple interest rate is 12%. How much should be invested in each account so that the total annual interest earned is $900?

2. An investment club invested a part of $14,000 in a 9% annual simple interest account and the remainder in a 15% annual simple interest account. The amount of interest earned for one year was $1800. How much was invested in each account?

3. An investment of $6000 is made at an annual simple interest rate of 14%. How much additional money must be invested at an annual simple interest rate of 10% so that the total interest earned is 12% of the total investment?

4. A total of $12,000 is invested into two simple interest accounts. On one account the annual simple interest rate is 12%, while on the second account the annual interest rate is 18%. How much should be invested in each account so that the interest earned is the same?

5. A total of $7000 is invested into two simple interest accounts. On one account the annual simple interest rate is 15%, while on the second account the annual simple interest rate is 20%. How much should be invested in each account so that the interest earned by each account is the same?

6. An accountant deposited an amount of money into a 10% annual simple interest account. Another deposit, $4000 more than the first, was placed in a 16% annual interest account. The total interest earned on both investments for one year was $2200. How much money was deposited into the 10% account?

7. An investment counselor for a corporation invested 70% of the company's investment account in 13% short term certificates. The remainder was invested in 12% corporate bonds. The annual interest earned from the two investments was $261,000. What was the total amount invested?

8. An investment of $20,000 is made in a 10.5% simple interest account. How much additional money must be deposited into an 8% simple interest account so that the total interest earned on both accounts is 9% of the total investment?

1. ______________

2. ______________

3. ______________

4. ______________

5. ______________

6. ______________

7. ______________

8. ______________

Name Score

Solve.

1. The owner of a delicatessen mixed coffee which cost \$4.50 per pound with coffee which cost \$3.00 per pound. How many pounds of each was used to make a 20 pound blend to sell for \$3.90 per pound?

2. How many pounds of cheese which cost \$3.80 per pound must be mixed with 16 lb of cheese which cost \$2.40 per pound to make a graded cheese topping which costs \$3.00 per pound?

1. ______________

2. ______________

3. A grocer mixed two varieties of tea, one that cost \$3 per kg with another that cost \$4 per kg. How many kilograms of each should be used to make a 30 kg blend to sell for \$3.40 per kilogram?

4. How many ounces of pure gold which cost \$360 an ounce must be mixed 30 oz of an alloy which cost \$240 an ounce to make an alloy which sells for \$320 an ounce?

3. ______________

4. ______________

5. A gardener has 10 lb of grass seed which cost \$0.80 per pound. How many pounds of a grass seed which cost \$1.20 per pound should be mixed with the 10 lb of grass seed to produce a mixture which sells for \$1.10 per pound?

6. Find the selling price per pound of a mixture made from 20 lb of chocolate which cost \$4.20 per pound and 25 lb of chocolate which cost \$2.40 per pound.

5. ______________

6. ______________

7. A grocer combined peanuts which cost \$2.60 per kilogram with cashews which cost \$3.20 per kilogram. How many kilograms of each were used to make a 45-kilogram mixture to sell for \$3.00 per kilogram?

8. Find the selling price per liter of a mixture made from 70 L of cranberry juice which cost \$1.20 per liter and 130 L of apple juice which cost \$0.80 per liter.

7. ______________

8. ______________

9. How many bushels of soybeans which cost \$6.00 per bushel must be mixed with 1500 bushels of corn which cost \$3.50 per bushel to make a mixture which cost \$5.00 per bushel?

10. Find the selling price per ounce of face cream mixture made from 80 oz of face cream which cost \$4.20 per ounce and 120 oz of face cream which cost \$2.60 per ounce.

9. ______________

10. ______________

Name Score

Solve.

1. A chemist has some 12% hydrogen peroxide solution and some 9% hydrogen peroxide solution. How many milliliters of each should be mixed to make a 450 milliliter solution which is 10% hydrogen peroxide?

2. How many grams of pure acid must be added to 60 g of a 15% acid solution to make a solution which is 40% acid?

1. ______________

2. ______________

3. How many ounces of pure water must be added to 75 oz of a 20% salt solution to make a salt solution which is 15% salt?

4. A syrup manufacturer has some pure maple syrup and some which is 80% maple syrup. How many liters of each should be mixed to get 120 L which is 95% maple syrup?

3. ______________

4. ______________

5. A butcher has some burger which is 15% fat and some which is 25% fat. How may pounds of each should be mixed to make 60 lb of hamburger which is 18% fat?

6. A 110-pound bag of animal feed is 30% oats. How many pounds of oats must be added to this feed to produce a mixture which is 45% oats?

5. ______________

6. ______________

7. Forty-five ounces of pure grapefruit juice is added to 30 oz of a fruit punch which is 15% grapefruit juice. What is the percent concentration of grapefruit juice in the resulting mixture?

8. A silversmith mixes 60 g of one alloy which is 50% silver with 200 g of another silver alloy. The resulting alloy is 65% silver. Find the percent of silver in the 200 gram alloy.

7. ______________

8. ______________

9. A clothing manufacturer has some fiber which is 25% polyester and some which is 55% polyester. How many pounds of each fiber should be woven together to product 300 lb of a fabric which is 35% polyester?

10. A baker mixed some flour which was 30% wheat with 100 lb of flour which was 45% wheat to make a mixture which is 36% wheat. How many pounds of the 30% wheat flour were used?

9. ______________

10. ______________

Name Score

Solve.

1. Two small planes start from the same point and fly in opposite directions. The first plane is 30 mph slower than the second plane. In three hours the planes are 1170 mi apart. Find the rate of each plane.

2. Two trains traveling on parallel tracks start at the same time from two towns 300 miles apart and travel toward each other. One of the trains travels 20 mph faster than the other. If they meet in 3 h, how fast is each traveling?

3. A family drove to a beach resort at an average speed of 44 mph and later returned over the same road at an average speed of 55 mph. Find the distance to the resort if the total driving time was 9 h.

4. Two cars, one traveling 12 mph faster than the other, start at the same time from the same point and travel in opposite directions. In four hours they are 384 mi apart. Find the rate of each car.

5. A motorboat leaves a harbor and travels at an average speed of 10 mph toward a small island. Three hours later a speed boat leaves the same harbor and travels at a rate of 18 mph toward the same island. In how many hours after the speed boat leaves will the speed boat be alongside the motorboat?

6. A pilot flew to Newark and back. The average speed going was 320 mph and the average speed returning was 480 mph. Find the distance to Newark if the total flying time was 5 h.

7. On a 300-mile trip, a car traveled at an average speed of 55 mph and then reduced the speed to 40 mph for the remainder of the trip. The trip took a total of 6 h. For how long did the car travel at each speed?

8. An executive flew in a helicopter to the airport to board a plane. The helicopter's flying speed was 120 mph and the airplane's flying speed was 600 mph. The entire trip was 2160 mi and took 4 h. How far did the executive fly in the helicopter?

9. Two cyclists start at the same time from opposite ends of a course which is 42 mi long. One cyclist is riding at 16 mph and the second is riding at 12 mph. How long after they begin will they meet?

10. Two planes start at the same time from the same point and fly in opposite directions. The first plane is flying 30 mph faster than the second plane. In four hours they are 2880 mi apart. Find the rate of each plane.

1. ____________
2. ____________
3. ____________
4. ____________
5. ____________
6. ____________
7. ____________
8. ____________
9. ____________
10. ____________

Name Score

Solve.

1. Six less than the product of five and a number is less than four plus the product of four and the number. Find the largest integer that will satisfy the inequality.

2. A student must receive 360 points out of 400 points on four tests to receive an A in an accounting course. One student's results on the first three tests were 82, 96, and 89. What scores on the last test will enable the student to receive an A in the course?

1. ____________

2. ____________

3. A sales manager for a computer company receives payments of $1450, $1870, and $1160 in commissions during a three-month period. What is the minimum commission the sales manager must earn during the fourth month in order to earn a minimum of $5500 during the four-month period?

4. An electrician's annual income for the last four years was $38,630, $30,960, $46,930, and $27,500. What is the minimum annual income the electrician must earn this year in order to have earned $180,000 during the five-year period?

3. ____________

4. ____________

5. Four times the sum of a number and five is less than two times the number. Find the largest integer that will satisfy the inequality.

6. A rectangle is 9 ft wide and $2x - 3$ ft long. Express as an integer the minimum length of the rectangle when the area is greater than 207 ft^2. (The area of a rectangle is equal to its length times its width.)

5. ____________

6. ____________

7. Company A rents cars for $10 a day and 10¢ for every mile driven. Company B rents cars for $28 a day with unlimited mileage. What is the maximum number of miles per day you can drive a Company A car if it is to cost you less than a Company B car?

8. Company A rents cars for $15 a day and 15¢ for every mile driven. Company B rents cars for $33 a day with unlimited mileage. You want to rent a car for one week. What is the maximum number of miles you can drive a Company A car if it is to cost you less than a Company B car?

7. ____________

8. ____________

9. An account executive earns $1200 per month plus 5% commission on the amount of sales. The executive's goal is to earn a minimum of $3400 per month. What amount of sales will enable the executive to earn $3400 or more per month?

10. An account executive earns $600 per month plus an 8% commission on the amount of sales. The executive's goal is to earn a minimum of $2800 per month. Find the minimum sales that will enable the executive to earn $2800 per month.

9. ____________

10. ____________

Name Score

Graph.

1. Graph the ordered pairs (–1, 2), (2, –4), (–1, 3), and (0, 1).

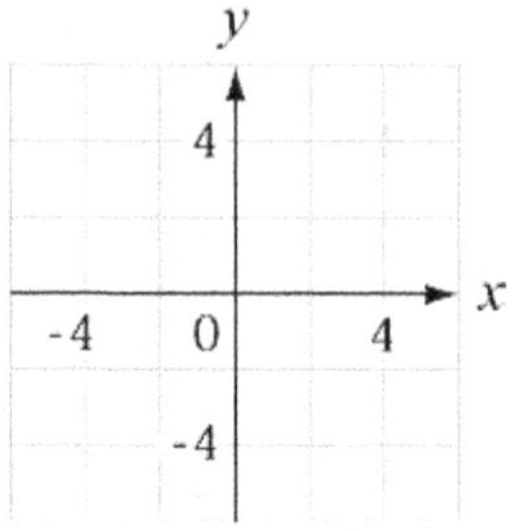

2. Graph the ordered pairs (4, –2), (–2, –2), (–2, 0), and (2, –1).

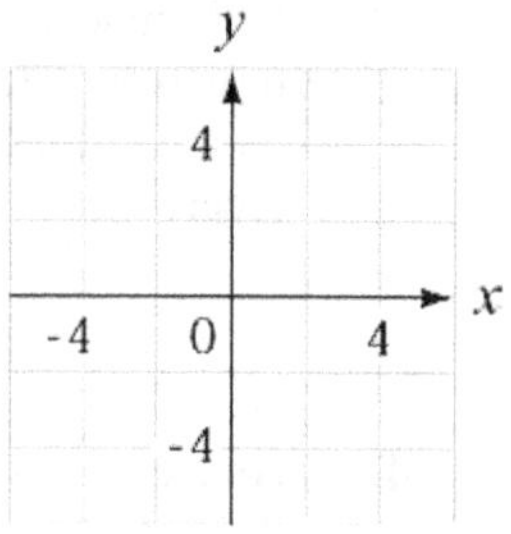

3. Graph the ordered pairs (1, 1), (1, –4), (–2, 1), and (1, –2).

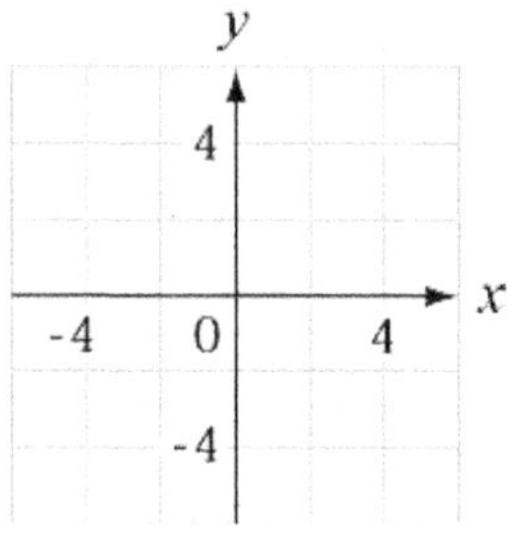

4. Graph the ordered pairs (–3, 4), (–2, 2), (2, –3), and (4, 0).

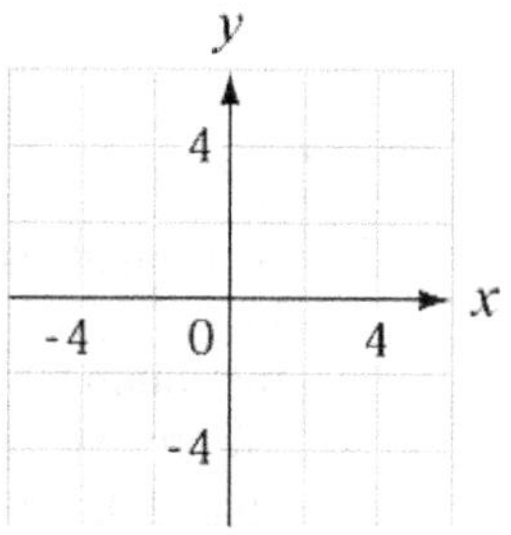

5. Graph the ordered pairs (2, –3), (–1, –2), (0, 1), and (4, 0).

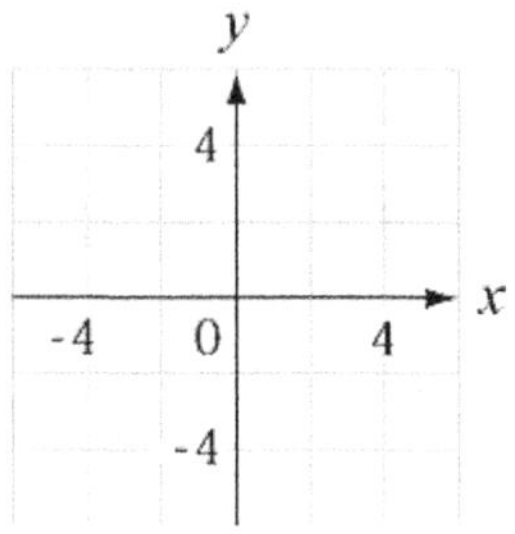

6. Graph the ordered pairs (4, 1), (–3, –2), (2, 2), and (0, –1).

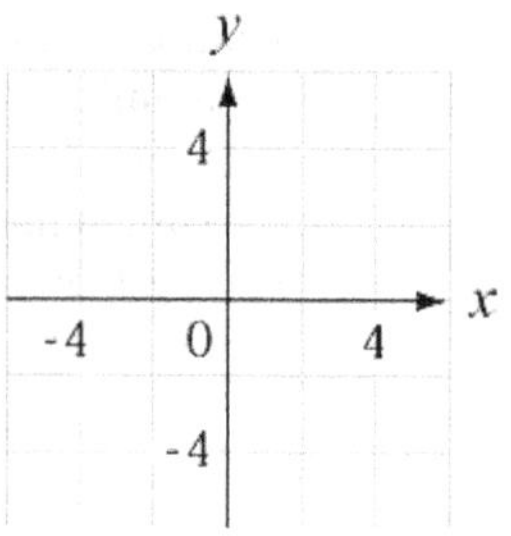

Name Score

Solve.

1. The table shows the number of U.S. households with a net worth of $1 million or more every three years from 1989 to 2001. Also shown is the millionaire households as a percent of all households in the United States. Graph the scatter diagram for these data.

Millionaire households (in millions), x	3.0	3.1	3.4	4.1	5.0
Percent of all households, y	3.2	3.2	3.4	4.0	4.7

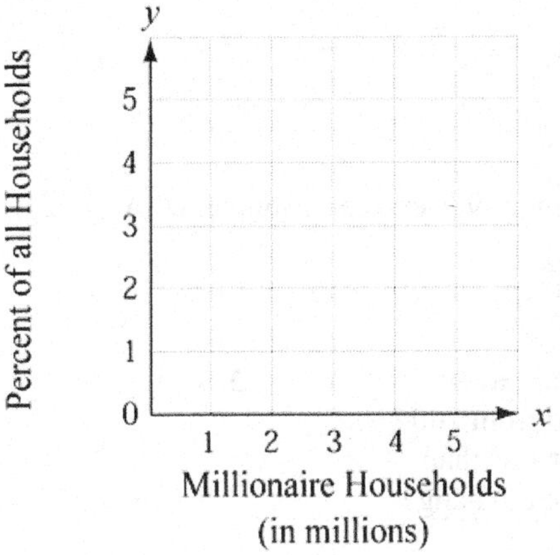

2. As shown in the table below, the percent of households with telephones has increased over the years. Draw a scatter diagram for these data.

Year, x	'20	'30	'40	'50	'60	'70	'80	'90	'00
Percent, y	35	41	37	62	78	90	93	95	97

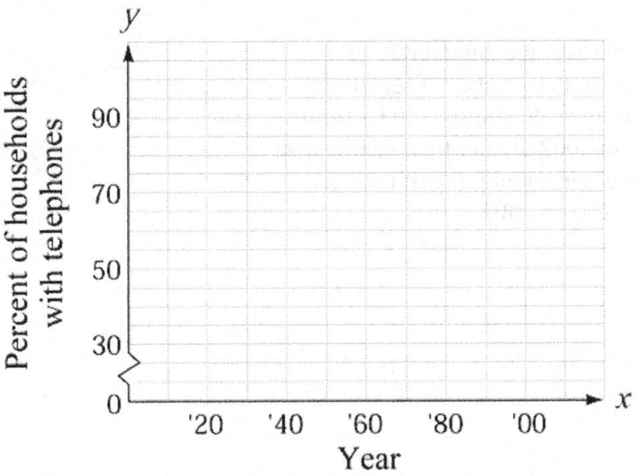

Name Score

Solve.

1. On February 21 in Sacramento, California, the temperature at 7 A.M. was 42°F. At 1 P.M. on the same day, the temperature was 66°F. Find the average rate of change in temperature per hour.

2. In 1970, the cost of a first class stamp was 6 cents. In 2000, the cost of a first class stamp was 33 cents. Find the average rate of change per year.

1. ____________

2. ____________

The table below shows the number of acres burned by wildfires through September 9 each year. (Source: USA Today)

Year	2002	2003	2004	2005	2006
Acres burned	6,420,000	3,060,000	7,610,000	7,930,000	8,650,000

3. What was the average annual rate of change in acres burned from 2004 to 2006?

4. Was the average annual rate of change in acres burned from 2004 to 2005 greater than or less than the average annual rate of change from 2005 to 2006?

3. ____________

4. ____________

The table below shows the number of registered lobbyists in Washington D.C. (Source: USA Today)

Year	2000	2001	2002	2003	2004	2005
Number registered	16,342	18,854	21,089	24,872	30,402	32,890

5. What was the average annual rate of change in the number of registered lobbyists in Washington D.C. from 2000 to 2004?

6. Was the average annual rate of change in the number of registered lobbyists in Washington D.C. from 2000 to 2002 greater than or less than the average annual rate of change from 2003 to 2005?

5. ____________

6. ____________

Name Score

1. Is (2, 3) a solution of $y = -x + 5$?

2. Is (1, –2) a solution of $y = x + 1$?

3. Is (–2, 4) a solution of $y = \frac{1}{2}x + 3$?

4. Is (2, –2) a solution of $y = -2x + 2$?

5. Is (3, 2) a solution of $y = x - 1$?

6. Is (–4, 2) a solution of $y = -\frac{3}{2}x + 2$?

7. Is (0, 3) a solution of $y = \frac{3}{5}x + 3$?

8. Is (–1, 0) a solution of $y = -x$?

9. Is (0, 0) a solution of $y = 2x - 3$?

10. Is (–1, –1) a solution of $y = x$?

11. Find the ordered pair solution of $y = 2x - 1$ corresponding to $x = 2$.

12. Find the ordered pair solution of $y = 3x + 2$ corresponding to $x = -3$.

13. Find the ordered pair solution of $y = \frac{1}{2}x - 2$ corresponding to $x = 4$.

14. Find the ordered pair solution of $y = \frac{2}{5}x - 3$ corresponding to $x = 10$.

15. Find the ordered pair solution of $y = -2x + 1$ corresponding to $x = -3$.

16. Find the ordered pair solution of $y = \frac{3}{5}x - 6$ corresponding to $x = 5$.

17. Find the ordered pair solution of $y = \frac{3}{4}x + 6$ corresponding to $x = -4$.

18. Find the ordered pair solution of $y = -\frac{1}{3}x - 1$ corresponding to $x = 6$.

1. ____________
2. ____________
3. ____________
4. ____________
5. ____________
6. ____________
7. ____________
8. ____________
9. ____________
10. ____________
11. ____________
12. ____________
13. ____________
14. ____________
15. ____________
16. ____________
17. ____________
18. ____________

Name Score

Graph.

1. $y = 2x - 5$

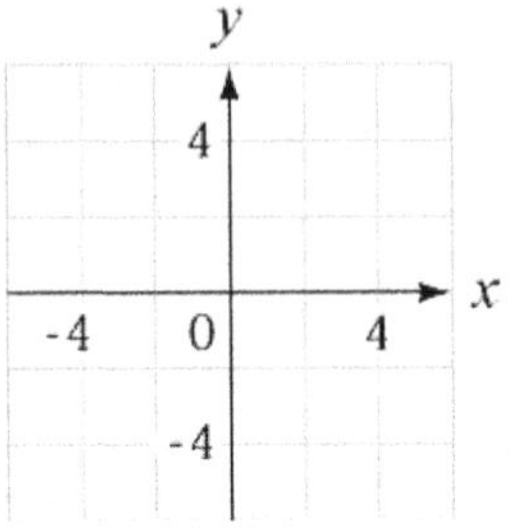

2. $y = -2x + 4$

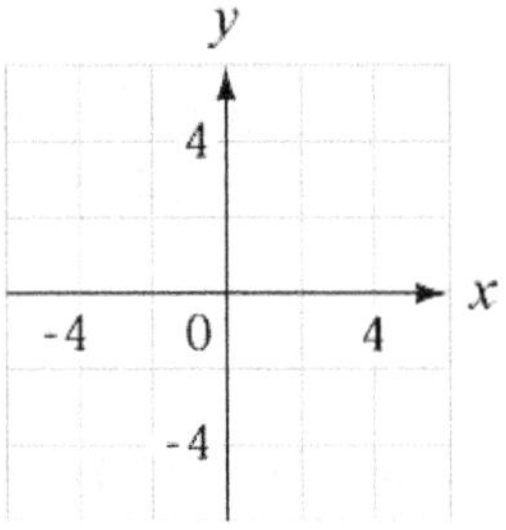

3. $y = \frac{1}{4}x$

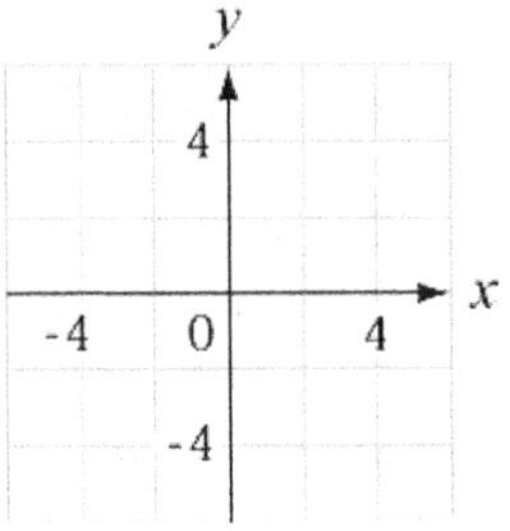

4. $y = -4x$

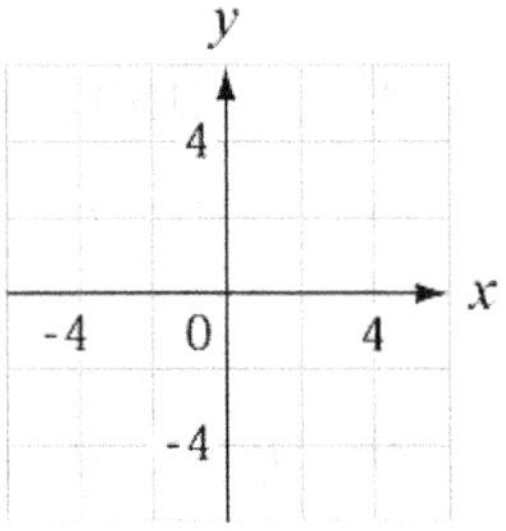

5. $y = \frac{1}{3}x + 1$

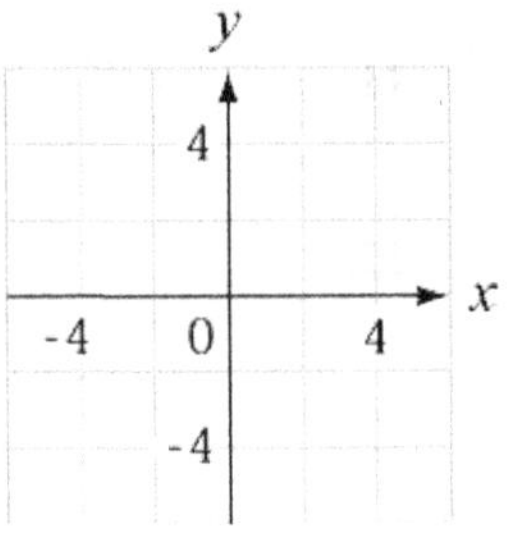

6. $y = \frac{2}{3}x - 1$

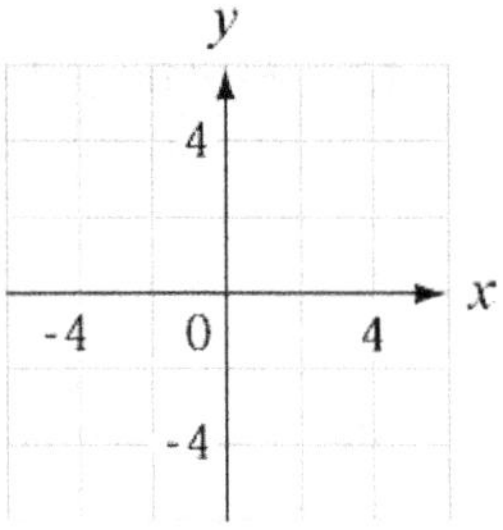

7. $y = 2x - 6$

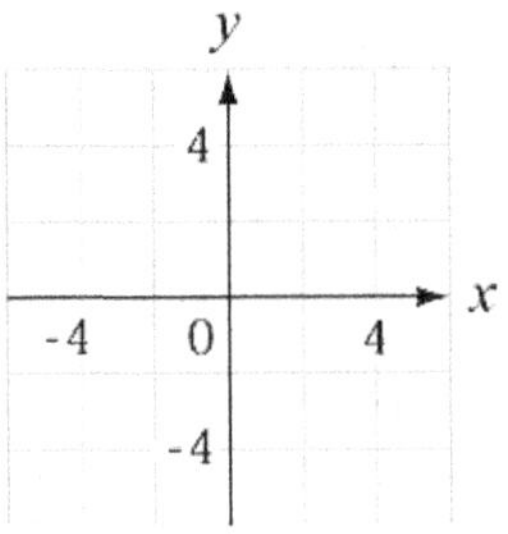

8. $y = -x + 1$

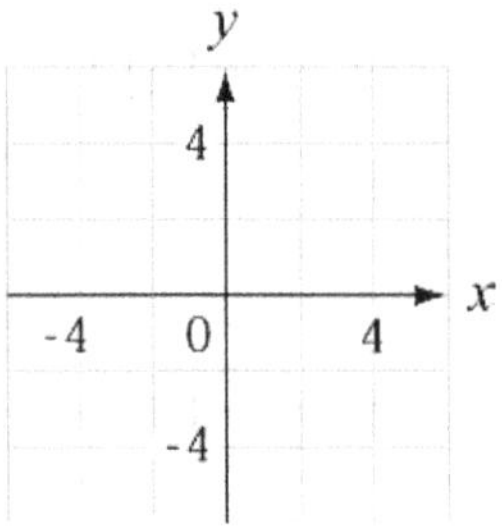

Name Score

Graph.

1. $3x - 2y = 6$

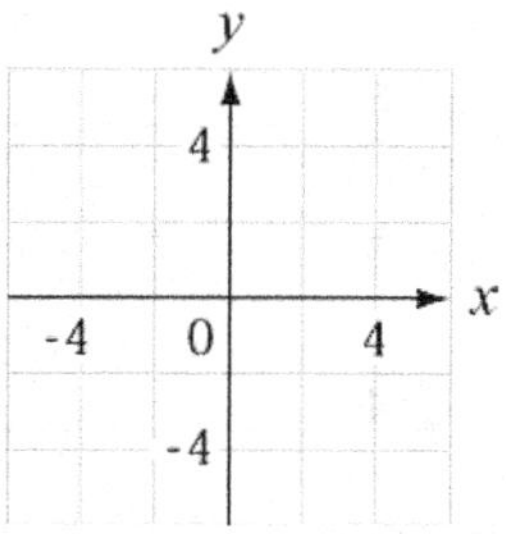

2. $y = 2$

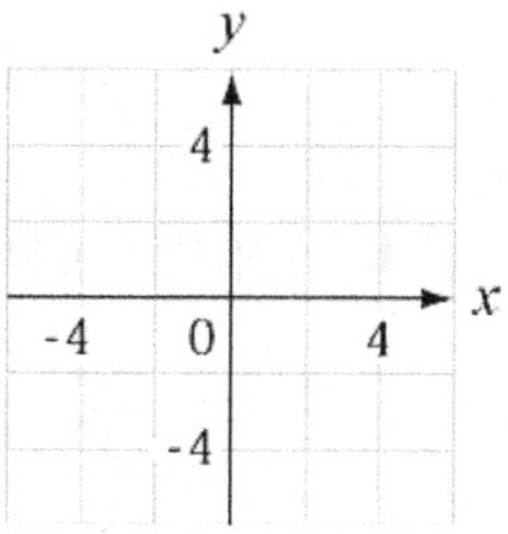

3. $x = -3$

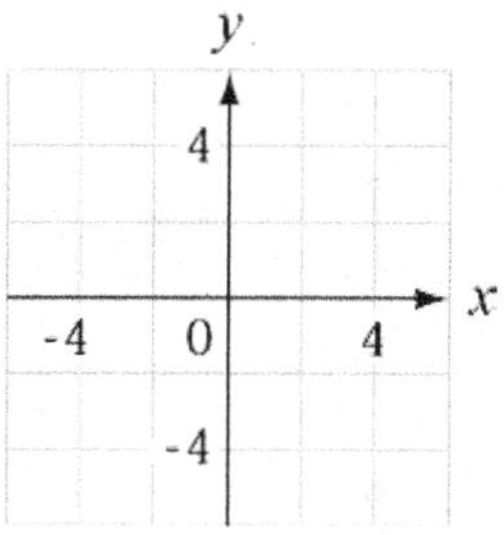

4. $x + 2y = 4$

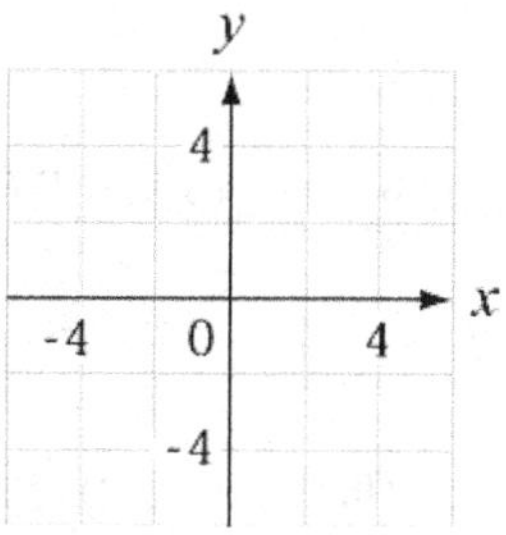

Find the *x*- and *y*-intercepts.

5. $2x + 5y = 10$

6. $y = 3x + 6$

7. $x - 4y = 8$

8. $y = 4x + 12$

9. $2x - 5y = 0$

10. $y = \frac{1}{3}x + 2$

5. ______________

6. ______________

7. ______________

8. ______________

9. ______________

10. ______________

Find the *x*- and *y*-intercepts and graph.

11. $x - 2y = 4$

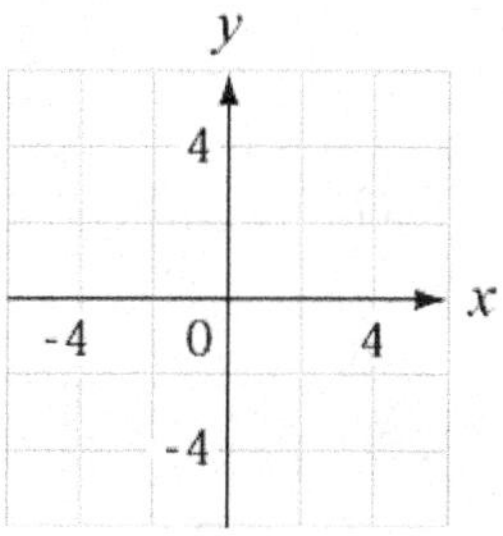

12. $4y = 3x + 12$

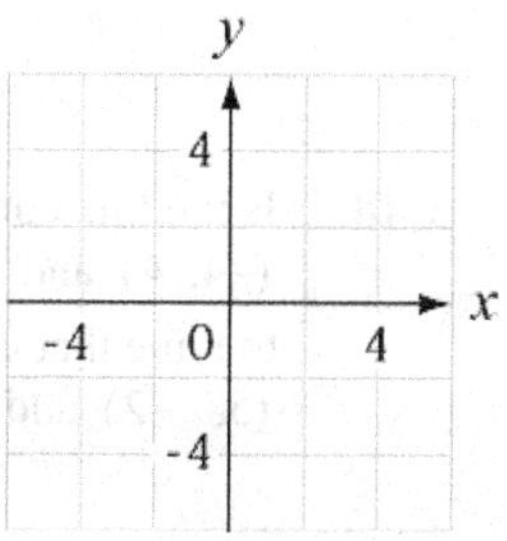

Name Score

Find the slope of the line containing the points.

1. $P_1(3, 1), P_2(2, 3)$

2. $P_1(3, 2), P_2(4, 5)$

3. $P_1(-1, 2), P_2(3, 3)$

4. $P_1(1, -1), P_2(3, -1)$

5. $P_1(0, 2), P_2(4, -2)$

6. $P_1(-2, 2), P_2(1, 3)$

7. $P_1(7, -2), P_2(3, -1)$

8. $P_1(0, 2), P_2(1, -2)$

9. $P_1(2, -3), P_2(2, 4)$

10. $P_1(-1, 3), P_2(-1, 1)$

11. $P_1(3, -3), P_2(4, -1)$

12. $P_1(4, 2), P_2(-1, 1)$

13. $P_1(1, -2), P_2(2, -1)$

14. $P_1(2, 0), P_2(3, -2)$

15. $P_1(-1, 3), P_2(-2, -2)$

16. $P_1(4, 1), P_2(3, -2)$

1. ____
2. ____
3. ____
4. ____
5. ____
6. ____
7. ____
8. ____
9. ____
10. ____
11. ____
12. ____
13. ____
14. ____
15. ____
16. ____

Solve.

17. Is the line containing the points $(5, -3)$ and $(3, -1)$ parallel to the line that contains the points $(4, -1)$ and $(2, 1)$?

18. Is the line containing the points $(-1, 4)$ and $(2, 2)$ parallel to the line that contains the points $(3, -2)$ and $(-2, -1)$?

17. ____
18. ____

Name Score

Graph by using the slope and *y*-intercept.

1. $y = 2x - 1$

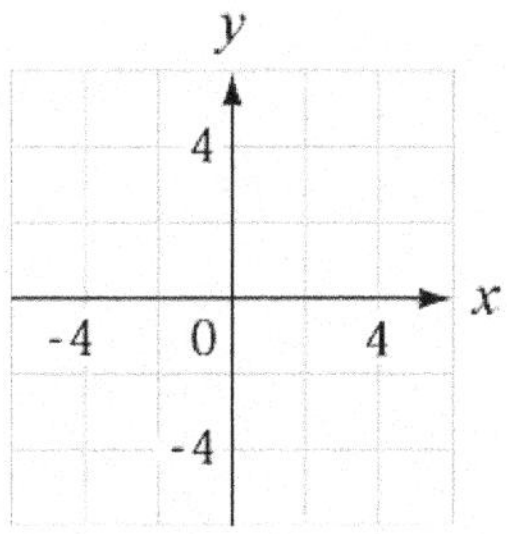

2. $2x - y = 3$

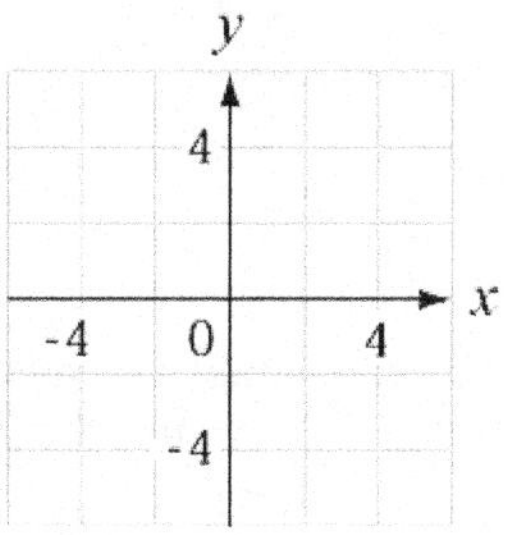

3. $x + 2y = -4$

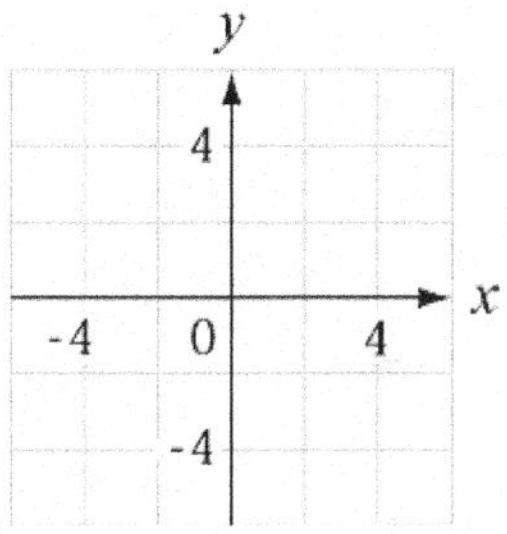

4. $x - 3y = 6$

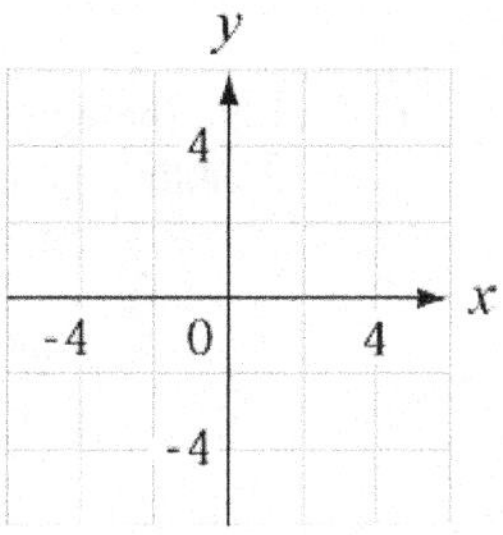

5. $y = -x + 2$

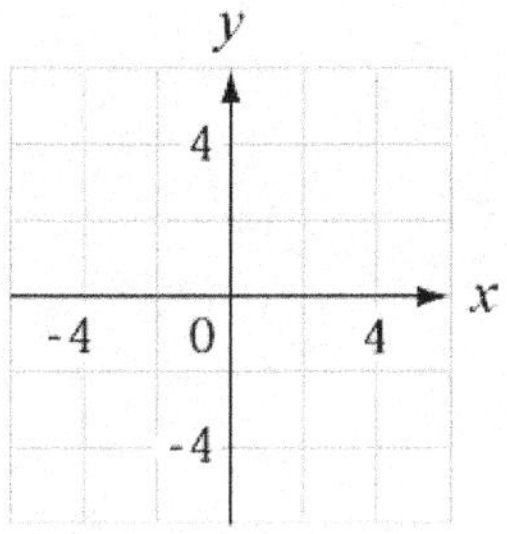

6. $y = -x - 4$

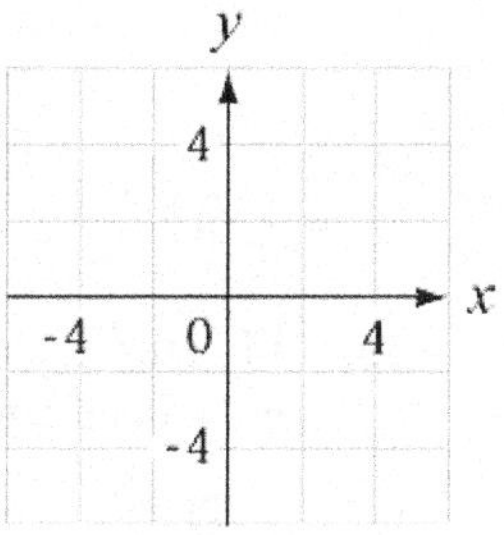

7. $y = -3x + 6$

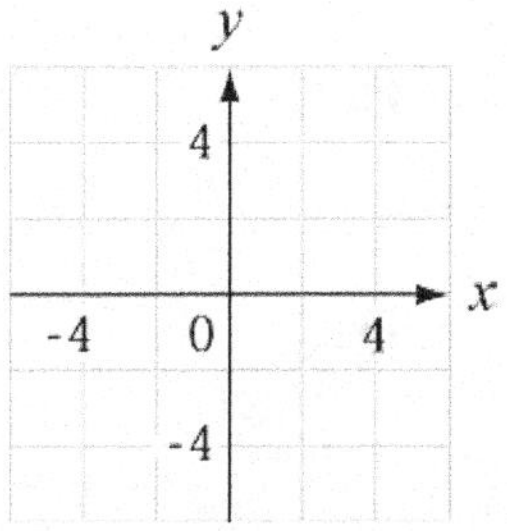

8. $2x - 5y = 10$

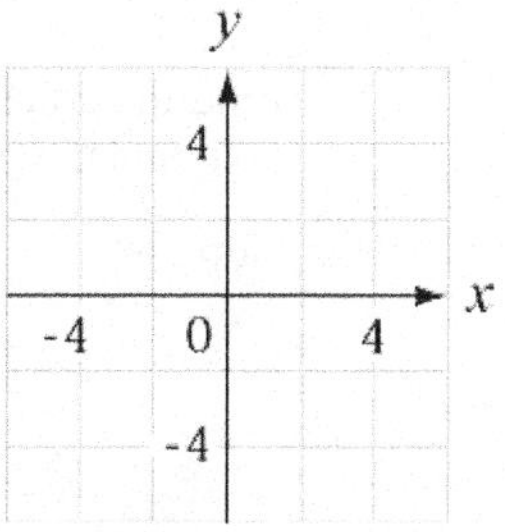

Name Score

Solve.

1. Find the equation of the line which contains the point (0, 1) and has slope 1.

2. Find the equation of the line which contains the point (–1, 0) and has slope –3.

3. Find the equation of the line which contains the point (2, 1) and has slope –2.

4. Find the equation of the line which contains the point (1, –2) and has slope 2.

5. Find the equation of the line which contains the point (2, 1) and has slope $\frac{1}{2}$.

6. Find the equation of the line which contains the point (–1, 3) and has slope $\frac{1}{3}$.

7. Find the equation of the line which contains the point (3, –3) and has slope $\frac{2}{3}$.

8. Find the equation of the line which contains the point (2, 4) and has slope $-\frac{1}{2}$.

9. Find the equation of the line which contains the point (5, –2) and has slope $-\frac{2}{5}$.

10. Find the equation of the line which contains the point (4, –1) and has slope $\frac{1}{4}$.

11. Find the equation of the line which contains the point (–1, 3) and has slope $-\frac{1}{3}$.

12. Find the equation of the line which contains the point (–2, 4) and has slope $-\frac{3}{4}$.

1. ____________

2. ____________

3. ____________

4. ____________

5. ____________

6. ____________

7. ____________

8. ____________

9. ____________

10. ____________

11. ____________

12. ____________

Name Score

Use the point-slope formula.

1. Find the equation of the line which contains the point (0, 1) and has slope 1.

2. Find the equation of the line which contains the point (–1, 0) and has slope –3.

3. Find the equation of the line which contains the point (2, 1) and has slope –2.

4. Find the equation of the line which contains the point (1, –2) and has slope 2.

5. Find the equation of the line which contains the point (2, 1) and has slope $\frac{1}{2}$.

6. Find the equation of the line which contains the point (–1, 3) and has slope $\frac{1}{3}$.

7. Find the equation of the line which contains the point (3, –3) and has slope $\frac{2}{3}$.

8. Find the equation of the line which contains the point (2, 4) and has slope $-\frac{1}{2}$.

9. Find the equation of the line which contains the point (5, –2) and has slope $-\frac{2}{5}$.

10. Find the equation of the line which contains the point (4, –1) and has slope $\frac{1}{4}$.

11. Find the equation of the line which contains the point (–1, 3) and has slope $-\frac{1}{3}$.

12. Find the equation of the line which contains the point (–2, 4) and has slope $-\frac{3}{4}$.

1. ____________
2. ____________
3. ____________
4. ____________
5. ____________
6. ____________
7. ____________
8. ____________
9. ____________
10. ____________
11. ____________
12. ____________

Name Score

Find the equation of the line through the given points.

1. $P_1(1, 4)$, $P_2(2, 3)$

2. $P_1(1, 3)$, $P_2(0, 4)$

3. $P_1(1, 2)$, $P_2(4, 4)$

4. $P_1(3, 1)$, $P_2(5, 2)$

5. $P_1(-2, 2)$, $P_2(1, 3)$

6. $P_1(-2, -2)$, $P_2(1, 1)$

7. $P_1(-2, -3)$, $P_2(2, 3)$

8. $P_1(-2, 1)$, $P_2(1, 5)$

9. $P_1(1, 0)$, $P_2(0, -2)$

10. $P_1(0, 3)$, $P_2(-1, 0)$

11. $P_1(0, 0)$, $P_2(2, 1)$

12. $P_1(1, -3)$, $P_2(0, 0)$

13. $P_1(1, 0)$, $P_2(-2, 2)$

14. $P_1(2, -4)$, $P_2(-1, 0)$

15. $P_1(3, -2)$, $P_2(-2, 4)$

16. $P_1(4, -5)$, $P_2(-3, 1)$

17. $P_1(-1, 3)$, $P_2(2, 0)$

18. $P_1(2, -2)$, $P_2(3, -2)$

1. ________
2. ________
3. ________
4. ________
5. ________
6. ________
7. ________
8. ________
9. ________
10. ________
11. ________
12. ________
13. ________
14. ________
15. ________
16. ________
17. ________
18. ________

Name Score

Find the domain and range of the function.

1. $\{(1, 1), (2, 3), (4, 3), (5, 7)\}$
2. $\{(1, 4), (6, 2), (3, 1), (5, 4)\}$

1. ____________
2. ____________

Find the domain and range of the relation. State whether or not the relation is a function.

3. $\{(1, 1), (2, 3), (4, 3), (5, 7)\}$
4. $\{(-3, -1), (-1, -2), (1, 5), (2, 6), (1, -7)\}$
5. $\{(-4, -2), (-2, -1), (0, 1), (3, 4)\}$
6. $\{(-3, 3), (3, -2), (-1, 4), (4, -1)\}$

3. ____________
4. ____________
5. ____________
6. ____________

Evaluate the function at the given value of *x*. Write an ordered pair that is an element of the function.

7. $f(x) = 2x + 1$; $x = -8$
8. $f(x) = 3x - 5$; $x = 0$
9. $f(x) = x^2 + 3x$; $x = -1$
10. $f(x) = 4x^2$; $x = -2$
11. $f(x) = \frac{5}{6}x + 2$; $x = -6$
12. $f(x) = 2x^2 + 7$; $x = -3$
13. $f(x) = \frac{1}{3}x - 2$; $x = 3$
14. $f(x) = 4x + 3$; $x = 10$
15. $f(x) = x^2 - 2x + 3$; $x = 0$

7. ____________
8. ____________
9. ____________
10. ____________
11. ____________
12. ____________
13. ____________
14. ____________
15. ____________

Find the range of the function defined by the equation and the given domain.

16. $f(x) = 4x + 3$
domain = $\{-1, 1, 3\}$
17. $g(x) = \frac{3}{4 - x}$
domain = $\{-1, 0, 3\}$
18. $f(x) = x^2 - 3x + 1$
domain = $\{-2, 1, 4\}$
19. $f(x) = \frac{3}{2}x - 1$
domain = $\{-2, 0, 4\}$

16. ____________
17. ____________
18. ____________
19. ____________

Name Score

Graph.

1. $y = 3x - 3$

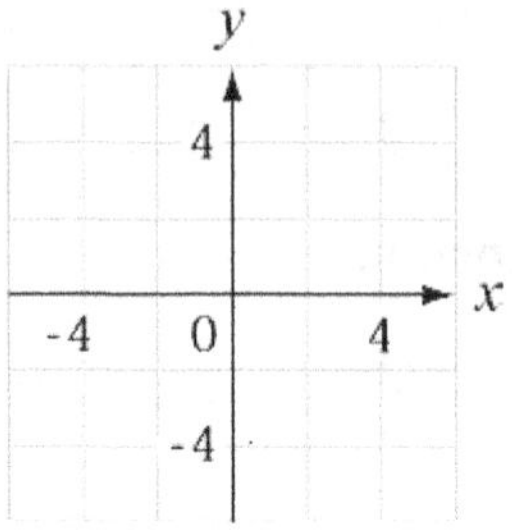

2. $y = \frac{4}{3}x$

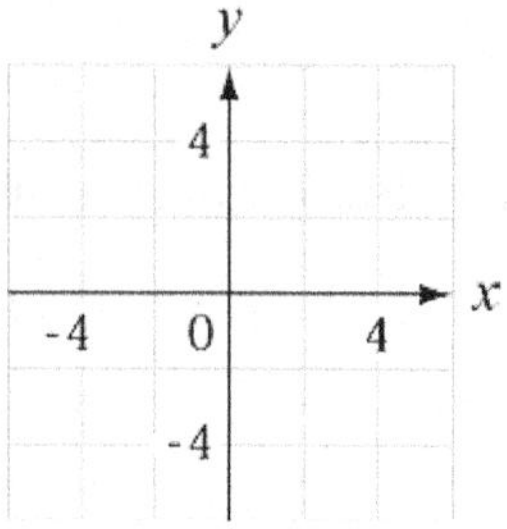

3. $y = -\frac{1}{3}x + 2$

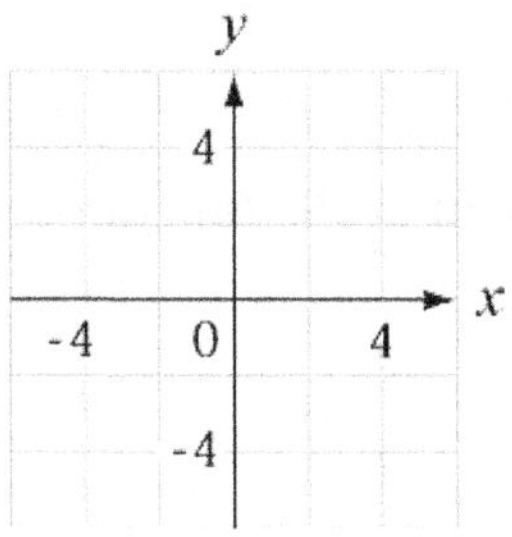

4. $y = \frac{1}{4}x - 1$

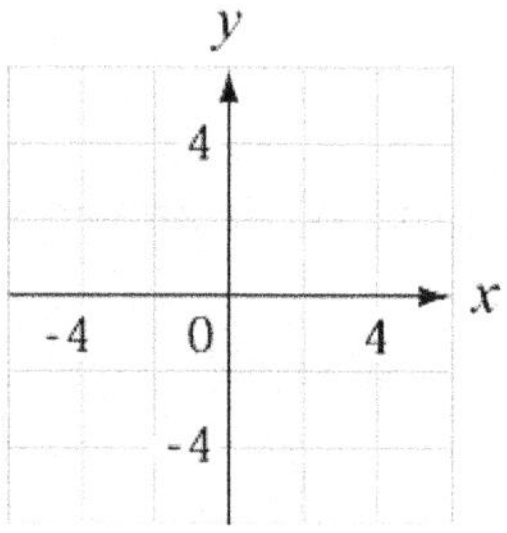

Solve.

5. The monthly cost for receiving messages from a telephone answering service is \$7.00 plus \$0.25 per message. The equation that describes the cost is $C = 0.25n + 7.00$, where n is the number of messages received.
a. Write the equation in functional notation.
b. Use the coordinate axes at the right to graph the equation for values of n between 0 and 40.
c. The point (36, 16) is on the graph. Write a sentence that describes the meaning of this ordered pair.

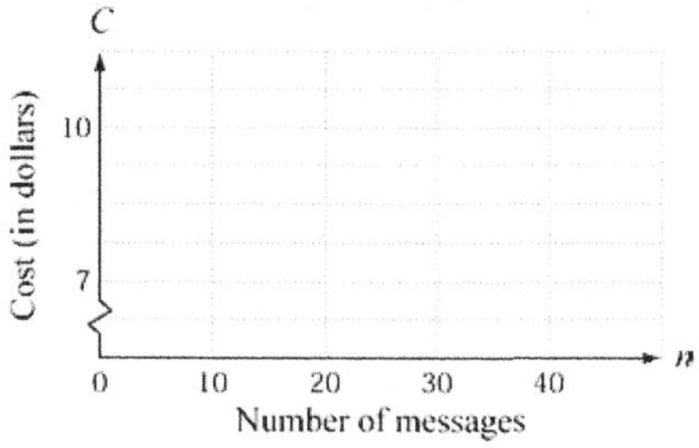

5. **a.** ______________

c. ______________

____

____

6. The cost of manufacturing snow boards is \$4500 for startup and \$75 per board manufactured. The equation that describes the cost of manufacturing n boards is $C = 75n + 4500$.
a. Write the equation in functional notation.
b. Use the coordinate axes at the right to graph the equation for values of n between 0 and 100.
c. The point (60, 9000) is on the graph. Write a sentence that describes the meaning of this ordered pair.

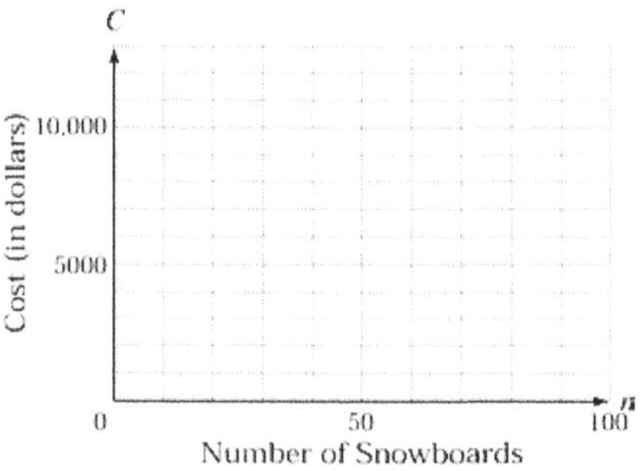

6. **a.** ______________

c. ______________

____

____

Name Score

Graph the solution set.

1. $y > -x + 2$

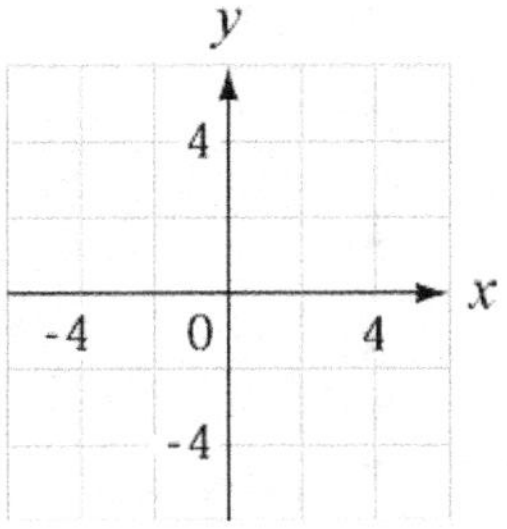

2. $y < x + 4$

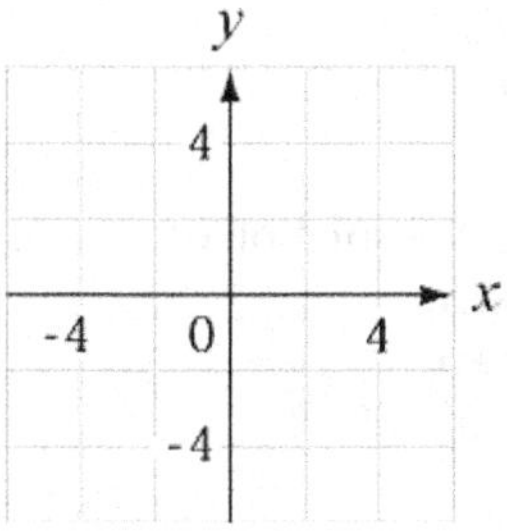

3. $2x - y < -4$

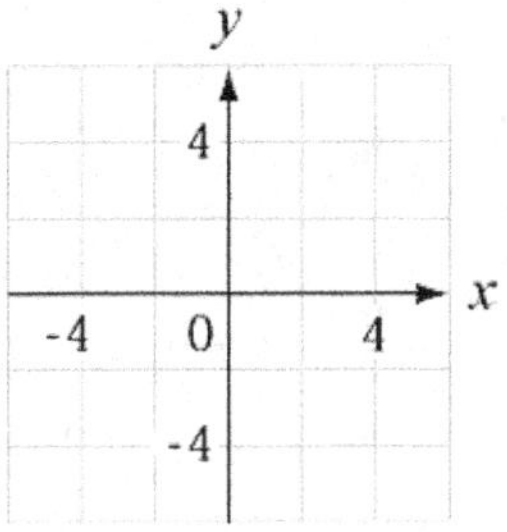

4. $3x - y < 6$

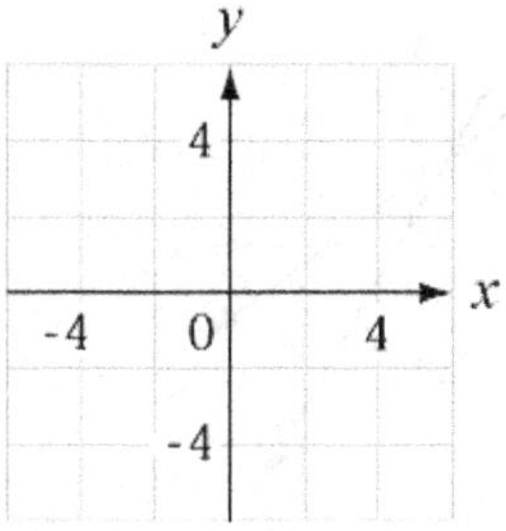

5. $y \geq -2x + 2$

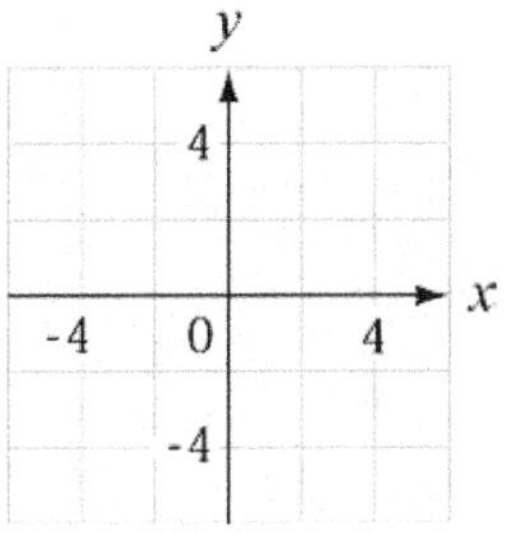

6. $y \geq -3x + 3$

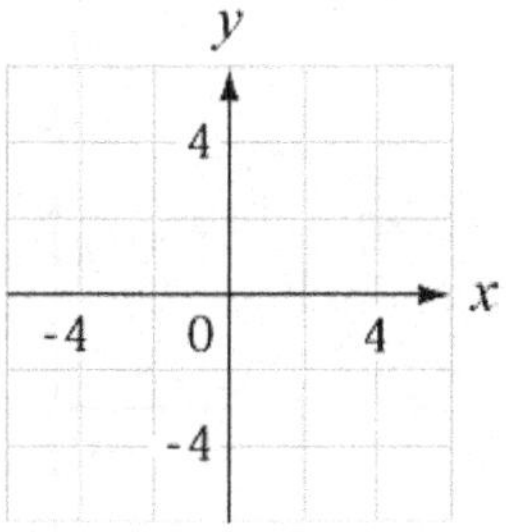

7. $y \leq -3$

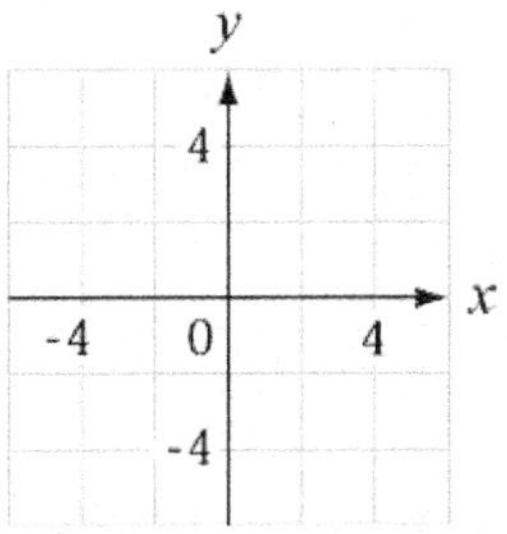

8. $x > 2$

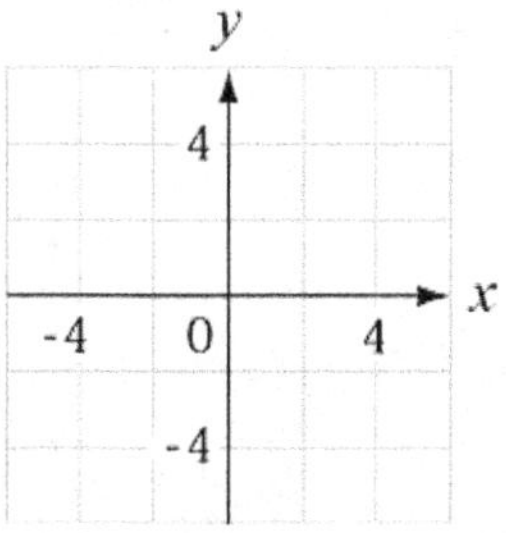

Name Score

Solve.

1. Is (1, 2) a solution of the system
$$2x+3y=8$$
$$2x-y=0\ ?$$

2. Is (3, –2) a solution of the system
$$x-2y=7$$
$$2x+y=-4\ ?$$

3. Is (–2, –2) a solution of the system
$$x-3y=4$$
$$3x+y=-8\ ?$$

4. Is (3, 2) a solution of the system
$$4x-y=10$$
$$2x+3y=0\ ?$$

5. Is (–1, 2) a solution of the system
$$3x+y=-5$$
$$2x+5y=8\ ?$$

6. Is (5, 0) a solution of the system
$$x-4y=5$$
$$2x-3y=10\ ?$$

1. ____________

2. ____________

3. ____________

4. ____________

5. ____________

6. ____________

State whether the system of equations is independent, inconsistent or dependent.

7.

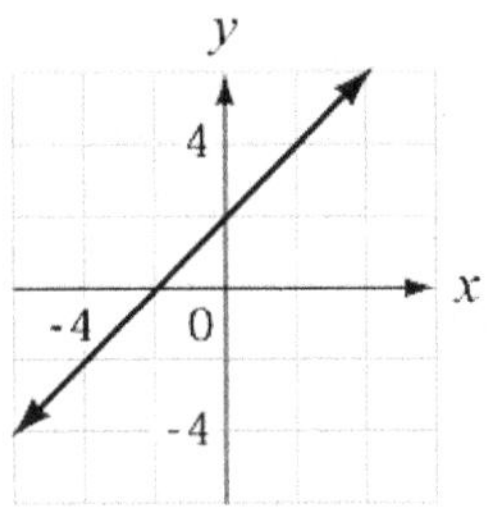

8.

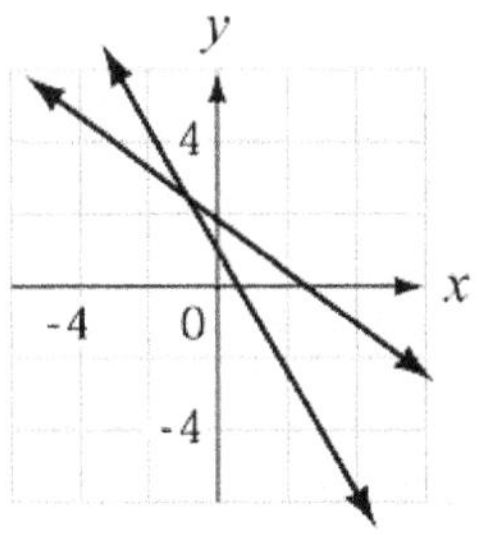

9.

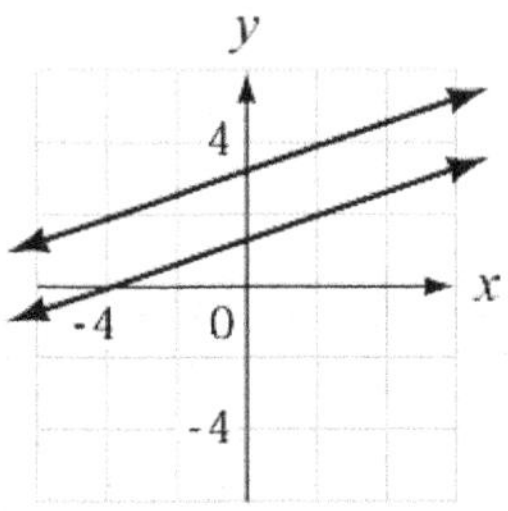

7. ____________

8. ____________

9. ____________

Solve by graphing.

10. $x-y=2$
$x+y=6$

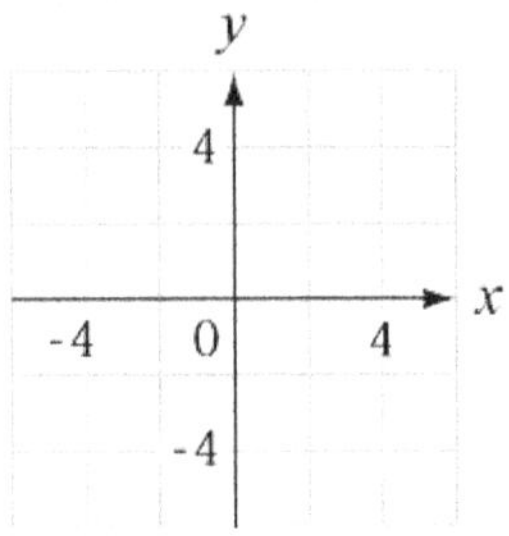

11. $x+y=3$
$2x+y=1$

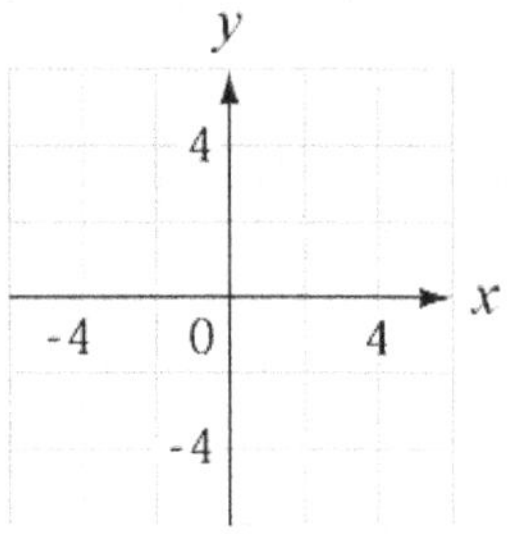

12. $x=3$
$2x-y=4$

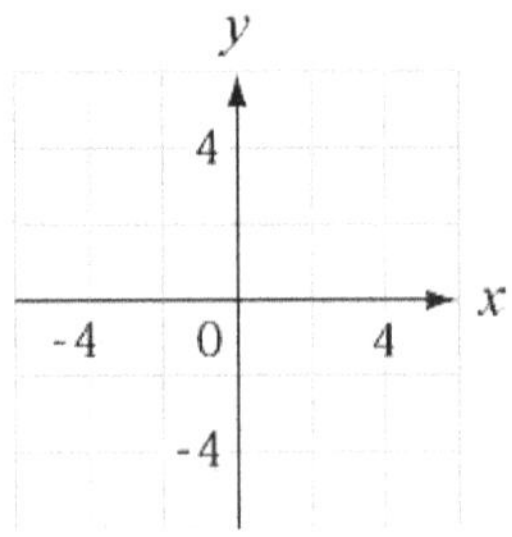

13. $x=2$
$y=-1$

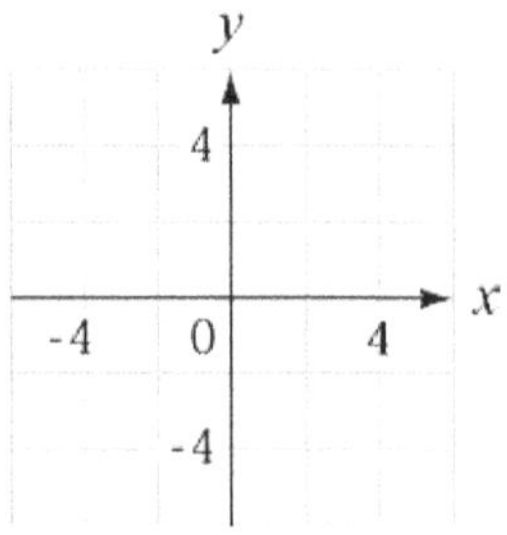

Name Score

Solve by substitution.

1. $3x+2y=5$
$x=1$

2. $y=2$
$4x-2y=8$

3. $y=x-2$
$x+y=6$

4. $y=x+3$
$x+y=5$

5. $x=y-1$
$x+2y=5$

6. $x=y+2$
$x+3y=14$

7. $2x+3y=6$
$y=x+7$

8. $2x+y=9$
$y=x-3$

9. $4x-y=9$
$y=2x-5$

10. $2x+y=5$
$4x-3y=-5$

11. $4x+y=3$
$5y+2x=-3$

12. $2x+y=0$
$3x-2y=7$

13. $x-4y=1$
$3x-5y=-4$

14. $y=2x+3$
$y=4x-1$

15. $y=-3x+4$
$y=-5x+10$

16. $x=6-2y$
$x=5y-8$

17. $y=2x-9$
$y=4x-23$

18. $-x+8y=-7$
$2x+5y=-7$

1. ____________
2. ____________
3. ____________
4. ____________
5. ____________
6. ____________
7. ____________
8. ____________
9. ____________
10. ____________
11. ____________
12. ____________
13. ____________
14. ____________
15. ____________
16. ____________
17. ____________
18. ____________

Name Score

Solve by addition method.

1. $x + y = 9$
   $x - y = 3$

2. $x + 4y = 6$
   $x - y = 1$

3. $x - 4y = 5$
   $x - 2y = 1$

4. $3x - 7y = 1$
   $2x - 7y = 3$

5. $2x - y = 8$
   $x - 8y = 4$

6. $3x - 5y = 5$
   $x + 2y = 9$

7. $x + 3y = 5$
   $3x + 2y = -6$

8. $7x - 5y = 11$
   $3x + y = -11$

9. $4x + 7y = 7$
   $x + 3y = -2$

10. $2x + 3y = 6$
    $x + 2y = 5$

11. $3x - 2y = -1$
    $2x - 3y = -4$

12. $4x - 3y = 10$
    $5x + 6y = -7$

13. $7x + 8y = -11$
    $3x + 4y = -7$

14. $5x - 2y = -8$
    $x + 3y = -5$

15. $3x + 4y = -7$
    $2x - 3y = 18$

16. $12x - 11y = -31$
    $5x + 7y = 45$

17. $5x + 9y = 6$
    $6x + 5y = 13$

18. $2x + 7y = 5$
    $3x - 5y = 23$

1. ________
2. ________
3. ________
4. ________
5. ________
6. ________
7. ________
8. ________
9. ________
10. ________
11. ________
12. ________
13. ________
14. ________
15. ________
16. ________
17. ________
18. ________

Name Score

Solve.

1. Paddling with the current, a canoeist can go 48 mi in 4 h. Against the current it takes 6 h to go the same distance. Find the rate of the canoeist in calm water and the rate of the current.

2. A pilot flying with the wind flew the 600 mi between two cities in 4 h. The return trip against the wind took 5 h. Find the rate of the plane in calm air and the rate of the wind.

3. A motorboat traveling with the current can go 45 mi in 3 h. Against the current it takes 5 h to go the same distance. Find the rate of the motorboat in still water and the rate of the current.

4. A plane flying with a tailwind flew 300 mi in 2 h. Against the wind, it took 3 h to fly the same distance. Find the rate of the plane in calm air and the rate of the wind.

5. A rowing team rowing with the current traveled 18 mi in 2 h. Against the current, the team rowed 12 mi in 2 h. Find the rate of the rowing team in calm water and the rate of the current.

6. A small jet plane flew 540 mi with the wind in 2 h. Against the wind, it took 3 h to travel the same distance. Find the rate of the plane in calm air and the rate of the wind.

7. A motorboat traveling with the current went 60 mi in 3 h. Against the current it took 5 h to travel the same distance. Find the rate of the motorboat in calm water and the rate of the current.

8. A jet plane flying with the wind went 2400 mi in 4 h. Against the wind, the plane could fly only 2000 mi in the same amount of time. Find the rate of the plane in calm air and the rate of the wind.

9. A crew can row 24 km downstream in 2 h. Rowing against the current, the team rowed 16 km in the same amount of time. Find the rowing rate of the crew in calm water and the rate of the current.

10. A plane flew 1600 km in 4 h traveling with the wind. Against the wind, the plane could only fly 1200 km in the same amount of time. Find the rate of the plane in calm air and the rate of the wind.

1. __________

2. __________

3. __________

4. __________

5. __________

6. __________

7. __________

8. __________

9. __________

10. __________

Name Score

Solve.

1. For the first performance of a play in a community theater, 60 reserved-seat tickets and 90 general-admission seat tickets were sold. The total receipts were $990. For the second performance, 80 reserved-seat tickets and 120 general-admission seat tickets were sold. The total receipts were $1320. Find the price of a reserved-seat ticket and the price of a general-admission ticket.

2. A video store received two shipments of video cassettes. The value of the first shipment, which contained 8 4-h and 16 8-h cassettes, was $224. The value of the second shipment, which contained 6 4-h and 18 8-h cassettes, was $228. Find the cost of a 4-h and the cost of an 8-h cassette.

1. ____________

2. ____________

3. Two coin banks contain only dimes and quarters. In the first bank, the total value of the coins is $6.50. In the second bank, there are 10 more dimes than in the first bank and one half as many quarters. The total value of the coins in the second bank is $5.00. Find the number of quarters in the first bank.

4. A citrus fruit grower purchased 30 orange trees and 20 lemon trees for $340. The next week, at the same prices, the grower bought 25 orange trees and 35 lemon trees for $430. Find the cost of an orange tree and a lemon tree.

3. ____________

4. ____________

5. A merchant mixed 20 lb of a cinnamon tea with 10 lb of a spice tea. The 30-pound mixture sells for $80. A second mixture included 12 lb of the cinnamon tea and 8 lb of the spice tea. The 20-pound mixture sells for $54. Find the cost per pound of the cinnamon tea and the spice tea.

6. A sheet metal shop ordered 30 lb of tin and 45 lb of zinc alloy for a total cost of $480. A second purchase, at the same price included 40 lb of tin and 60 lb of the zinc alloy. The total cost was $640. Find the cost per pound for the tin and the zinc alloy.

5. ____________

6. ____________

7. A restaurant manager buys 50 lb of hamburger and 80 lb of steak for a total cost of $300. A second purchase, at the same price, includes 100 lb of hamburger and 120 lb of steak. The total cost is $480. Find the cost per pound of the hamburger and the steak.

8. The total value of the nickels and quarters in a coin bank is $6.50. If the nickels were quarters and the quarters were nickels, the total value of the coins would be $8.50. Find the number of nickels and the number of quarters in the bank.

7. ____________

8. ____________

Name Score

Add. Use a vertical format.

1. $(x^2+5x)+(-2x^2-3x)$

2. $(2y^2-y)+(3y^2+4y)$

3. $(y^2+3y)+(-2y-5)$

4. $(2x^2+5x)+(2x-12)$

5. $(4x^2+7x)+(3x-8)$

6. $(y^2-2y)+(-3y-6)$

7. $(x^2+5x+10)+$
$(2x^2+3x+1)$

8. $(3x^2+4x+9)+$
(x^2-x-6)

9. $(2y^3-y^2-3)+$
$(-2y^3-y^2-6)$

10. $(4y^3+2y^2-2)+$
$(-3y^3-2y-1)$

11. $(3a^3-7a^2-3)+$
$(-3a^2+7a+3)$

12. $(3y^3+2y^2-y)+$
$(-y^3-5y-2)$

Add. Use a horizontal format.

13. $(3x^2+x)+(2x^2+5x)$

14. $(-2y^2+3y)+(3y^2+5y)$

15. $(5x^2-4xy)+$
$(2x^2+5xy-2y^2)$

16. $(6x^2-3xy)+$
$(x^2+7xy-3y^2)$

17. $(-5x^2-8x-2)+$
$(3x^2-x-5)$

18. $(a^2-6a+9)+$
$(3a^2+3a+6)$

19. $(-7x^2+6x+4)+$
$(2x^2-x-3)$

20. $(2a^2-9a+5)+$
$(3a^2+3a+6)$

21. $(-4x^2+3x+1)+$
$(-x^2-x-2)$

22. $(4x^3+6x-6)+$
$(9x^2-7x+4)$

23. $(2y^2-3y+9)+$
$(5y^3-5y)$

24. $(6x^3+6x-3)+$
(x^2-6x-5)

1. ________
2. ________
3. ________
4. ________
5. ________
6. ________
7. ________
8. ________
9. ________
10. ________
11. ________
12. ________
13. ________
14. ________
15. ________
16. ________
17. ________
18. ________
19. ________
20. ________
21. ________
22. ________
23. ________
24. ________

Name Score

Subtract. Use a vertical format.

1. $(x^2-3x)-(x^2-7x)$
2. $(y^2+2y)-(y^2+8y)$
3. $(2y^2-3y)-(y^2+6)$
4. $(x^2-3x+2)-$ (x^2+4x+5)
5. $(2x^2+3x-1)-$ (x^2+5x+8)
6. $(x^2+3x+4)-$ $(2x^2-9x-2)$
7. $(3x^3+6x+3)-$ $(-2x^2+3x+2)$
8. $(-3x^2-2x+3)-$ $(2x^3+x^2+3)$
9. $(3x^3+6x+3)-$ $(-2x^2+3x+2)$
10. $(-3x^2-2x+3)-$ $(2x^3+x^2+3)$
11. $(3y^3+2y^2-y)-$ $(-y^3-5y-2)$
12. $(2x^3-5x+6)-$ (x^3-x+7)

1. ____________
2. ____________
3. ____________
4. ____________
5. ____________
6. ____________
7. ____________
8. ____________
9. ____________
10. ____________
11. ____________
12. ____________

Subtract. Use a horizontal format.

13. $(y^2-6xy)-(y^2-3xy)$
14. $(x^2-2xy)-(-2x^2+3xy)$
15. $(2y^2-8xy)-(y^2-6xy)$
16. $(6y^2-y+6)-$ $(-3y^2-2y-1)$
17. $(3x^2+2x-2)-$ (x^2+5x-6)
18. $(2x^2+6x-1)-$ $(5x^2-x-2)$
19. $(-x^3+x^2-x)-$ $(-x^2-x-7)$
20. $(3x^3-2x-6)-$ $(2x^2+6x-1)$
21. $(2x^3+4x-7)-$ $(-x^2+x-13)$
22. $(3x^3+4x-4)-$ $(8x^2-6x-1)$
23. $(-4x^2+3x-1)-$ $(-x^2-x-2)$
24. $(a^2-7a+5)-$ $(2a^3-6a-1)$

13. ____________
14. ____________
15. ____________
16. ____________
17. ____________
18. ____________
19. ____________
20. ____________
21. ____________
22. ____________
23. ____________
24. ____________

Name Score

Simplify.

1. $(x)(3x)$
2. $(-2y)(y)$
3. $(2x)(3x)$
4. $(4y^3)(4y^2)$
5. $(-2a^2)(-4a^3)$
6. $(4a^6)(-2a^4)$
7. $(xy^2)(x^2y^3)$
8. $(-3x^3)(4x^2y)$
9. $(-3a^3)(3a^3b^2)$
10. $(ab^3)(a^2b^2)$
11. $(4xy^2)(-2x^2y^3)$
12. $(-5ab^2)(-a^2b^3)$
13. $(xy^2z)(x^2y^3)$
14. $(ab^2)(a^2bc^3)$
15. $(xy^2z)(x^3y^2)$
16. $(-ab^2)(a^3b^5)$
17. $(x^2y^3)(-x^2y)$
18. $(-4y^3z)(-7y^4z^3)$
19. $(2x^2y)(-3x^2y)$
20. $(-a^2bc^2)(ab^2c)(-abc^3)$
21. $(x^2y^2z)(x^2y^3)(y^3z^2)$
22. $(-2ab)(3a^2b^2)(-4a^3b^3)$
23. $(2x^2y^4)(4xy)(-3x^2y^2)$
24. $(2a^2b^2)(-3ab^2)(a^4b^2)$
25. $(x^2y^3)(-2xy)(-4x^2y^4)$
26. $(3a^2b)(-2a^3b^2)(a^5b)$
27. $(2ab^2)(-3abc^2)(5a^2c)$

1. ________
2. ________
3. ________
4. ________
5. ________
6. ________
7. ________
8. ________
9. ________
10. ________
11. ________
12. ________
13. ________
14. ________
15. ________
16. ________
17. ________
18. ________
19. ________
20. ________
21. ________
22. ________
23. ________
24. ________
25. ________
26. ________
27. ________

Name Score

Simplify.

1. $(x^2)^2$
2. $(x^2)^3$
3. $(y^2)^4$
4. $(x^5)^2$
5. $(y^4)^3$
6. $(-x^3)^2$
7. $(2x)^3$
8. $(3y)^2$
9. $(-2x)^3$
10. $(-3y^2)^2$
11. $(xy^2)^2$
12. $(x^2y^3)^4$
13. $(-2ab^2)^3$
14. $(x^2y^2)^5$
15. $(xy^3)^5$
16. $(x^4y^3)^4$
17. $(2x^2y^3)^2$
18. $(-a^2b^4)^3$
19. $(a^2b)^2(ab^2)^2$
20. $(-2y)(3y^2)^3$
21. $(a^2b)(ab)^2$
22. $(-3x)(-2x^2y^2)^2$
23. $(-2y)(-3x^2y)^2$
24. $(ab^2)(-2ab^3)^2$
25. $(-2a^2)(3ab^2)^3$
26. $(2a^2b)^3(-3b^2)$
27. $(-3ab^2)^3(-3ab)^2$

1. ____
2. ____
3. ____
4. ____
5. ____
6. ____
7. ____
8. ____
9. ____
10. ____
11. ____
12. ____
13. ____
14. ____
15. ____
16. ____
17. ____
18. ____
19. ____
20. ____
21. ____
22. ____
23. ____
24. ____
25. ____
26. ____
27. ____

Name Score

Multiply.

1. $x(x+1)$

2. $y(2-y)$

3. $-x(x+2)$

4. $2a(a-1)$

5. $3b(b+5)$

6. $-2x^2(x-1)$

7. $-4y^2(y+6)$

8. $-x^2(2x^2-3)$

9. $2x(5x^2-2x)$

10. $3y(2y-y^2)$

11. $(2x-3)x$

12. $(2x-1)3x$

13. $-x^2y(x-y^2)$

14. $-xy^2(2x-y)$

15. $x(x^2-2x+1)$

16. $x(x^2+3x-2)$

17. $-a(a^2-6a-1)$

18. $-b(2b^2+3b-6)$

19. $x^2(2x^2-3x-2)$

20. $y^3(-2y^2-3y+4)$

21. $2y^2(-2y^2-5y+8)$

22. $3x^2(4x^2-2x+7)$

23. $4x^2(5x^2-x-9)$

24. $5y^2(-y^2+3y-6)$

25. $ab(a^2-3ab-4b^2)$

26. $xy(x^2-2xy+2y^2)$

27. $ab(a^2+5ab-7b^2)$

1. ____________
2. ____________
3. ____________
4. ____________
5. ____________
6. ____________
7. ____________
8. ____________
9. ____________
10. ____________
11. ____________
12. ____________
13. ____________
14. ____________
15. ____________
16. ____________
17. ____________
18. ____________
19. ____________
20. ____________
21. ____________
22. ____________
23. ____________
24. ____________
25. ____________
26. ____________
27. ____________

Name Score

Multiply.

1. $(x^2+2x+1)(x+2)$

2. $(x^2-3x+5)(x-1)$

3. $(a^2-a+2)(a-2)$

4. $(x^2-4x-5)(x-3)$

5. $(a^2-5a+3)(a-1)$

6. $(x^2-5x-4)(3x-1)$

7. $(-2b^2-b+2)(b-4)$

8. $(-a^2+4a-3)(2a+3)$

9. $(-x^2+6x-3)(2x-3)$

10. $(-a^2-a+2)(3a+1)$

11. $(x^2-4)(x+1)$

12. $(y^2+3y)(y-3)$

13. $(x^2+3)(x-5)$

14. $(y^2-y)(2y+4)$

15. $(x^3-2x+1)(x-3)$

16. $(y^3+2y^2-5)(2y-3)$

17. $(4y^2+7y-1)(2y-7)$

18. $(3y^2+y-3)(3y-2)$

19. $(3a^3-2a+3)(a-2)$

20. $(2y^2+4y-2)(2y-1)$

21. $(4a^3-4a+1)(a-3)$

22. $(2b^3-3b^2+6)(3b-1)$

23. $(y^3+y^2-2y+2)(y+1)$

24. $(a^3-2a^2+3a-1)(3a-2)$

1. ______
2. ______
3. ______
4. ______
5. ______
6. ______
7. ______
8. ______
9. ______
10. ______
11. ______
12. ______
13. ______
14. ______
15. ______
16. ______
17. ______
18. ______
19. ______
20. ______
21. ______
22. ______
23. ______
24. ______

Name Score

Multiply.

1. $(y+2)(y+3)$
2. $(a-2)(a+5)$
3. $(b-5)(b+4)$
4. $(y+2)(y-7)$
5. $(x+9)(x-4)$
6. $(y-6)(y-2)$
7. $(a-2)(a-8)$
8. $(x+11)(x-3)$
9. $(2x+1)(x+6)$
10. $(y+1)(3y+2)$
11. $(2x-3)(x+3)$
12. $(5x-2)(x+3)$
13. $(2x-1)(3x-5)$
14. $(2y-9)(y+1)$
15. $(4y-7)(y+2)$
16. $(5a+2)(6a+1)$
17. $(6a-13)(2a-5)$
18. $(5a-9)(2a-7)$
19. $(3b+11)(5b-4)$
20. $(3a+10)(4a-3)$
21. $(x+y)(x+2y)$
22. $(2a+b)(a+2b)$
23. $(2x-3y)(x-y)$
24. $(a-3b)(2a+3b)$
25. $(4a-b)(2a+5b)$
26. $(3x-5y)(3x+2y)$
27. $(5x+2y)(6x+y)$

1. ______
2. ______
3. ______
4. ______
5. ______
6. ______
7. ______
8. ______
9. ______
10. ______
11. ______
12. ______
13. ______
14. ______
15. ______
16. ______
17. ______
18. ______
19. ______
20. ______
21. ______
22. ______
23. ______
24. ______
25. ______
26. ______
27. ______

Name Score

Simplify.

1. $(x+1)(x-1)$
2. $(x-3)(x+3)$
3. $(x-7)(x+7)$
4. $(2x-1)(2x+1)$
5. $(3x-1)(3x+1)$
6. $(x+5)^2$
7. $(y-4)^2$
8. $(3y-1)^2$
9. $(x-1)^2$
10. $(x+7)^2$
11. $(x+9)^2$
12. $(x-y)^2$
13. $(2a-5)^2$
14. $(5x-4)^2$
15. $(3x-7)^2$
16. $(3a-5)(3a+5)$
17. $(2x+5)^2$
18. $(9x-2)^2$
19. $(a-2b)^2$
20. $(x+2y)^2$
21. $(5x-6)(5x+6)$
22. $(b-6a)(b+6a)$
23. $(x+5y)^2$
24. $(2-7y)^2$
25. $(3-5y)(3+5y)$
26. $(4x-1)(4x+1)$
27. $(x+6y)^2$

1. ____________
2. ____________
3. ____________
4. ____________
5. ____________
6. ____________
7. ____________
8. ____________
9. ____________
10. ____________
11. ____________
12. ____________
13. ____________
14. ____________
15. ____________
16. ____________
17. ____________
18. ____________
19. ____________
20. ____________
21. ____________
22. ____________
23. ____________
24. ____________
25. ____________
26. ____________
27. ____________

Name Score

Solve.

1. The length of a rectangle is $3x$ m. The width is $(3x - 1)$ m. Find the area of the rectangle in terms of the variable x.

2. The width of a rectangle is $(x - 2)$ in. The length is $(3x + 2)$ in. Find the area of the rectangle in terms of the variable x.

3. The length of a rectangle is $(3x + 1)$ ft. The width is $(2x - 1)$ ft. Find the area of the rectangle in terms of the variable x.

4. The length of a side of a square is $(x + 3)$ cm. Use the equation $A = s^2$, where s is the length of a side of a square, to find the area of the square in terms of the variable x.

5. The length of a side of a square is $(x - 8)$ km. Use the equation $A = s^2$, where s is the length of a side of a square, to find the area of the square in terms of the variable x.

6. The length of a side of a square is $(2x + 1)$ in. Find the area of the square in terms of the variable x.

7. The length of a side of a square is $(3x - 4)$ ft. Find the area of the square in terms of the variable x.

8. The radius of a circle is $(x - 3)$ m. Use the equation $A = \pi r^2$, where r is the radius, to find the area of the circle in terms of the variable x. Leave the answer in terms of π.

9. The radius of a circle is $(x + 6)$ in. Use the equation $A = \pi r^2$, where r is the radius, to find the area of the circle in terms of the variable x. Leave the answer in terms of π.

10. The radius of a circle is $(2x + 1)$ ft. Use the equation $A = \pi r^2$, where r is the radius, to find the area of the circle in terms of the variable x. Leave the answer in terms of π.

1. ______________

2. ______________

3. ______________

4. ______________

5. ______________

6. ______________

7. ______________

8. ______________

9. ______________

10. ______________

Name Score

Write with a positive or zero exponent. Then evaluate.

1. 6^{-2}

2. $\frac{2^{-4}}{2^2}$

3. $\frac{5^{-2}}{5}$

1. ____________
2. ____________
3. ____________

Simplify.

4. a^{-5}

5. xy^{-2}

6. x^3x^{-3}

7. $\frac{a^{-7}}{a^7}$

8. $\frac{x^{-1}y^2}{x}$

9. $(x^2)^{-4}$

10. $(a^{-1}b^2)^0$

11. $\frac{6x^2}{2x}$

12. $\frac{18x^5}{-6x^2}$

13. $\frac{a^7b^9}{a^9b^3}$

14. $\frac{x^8y^6}{x^4y^3}$

15. $\frac{(3y)^2}{9y}$

16. $\frac{(5a)^3}{25a^2}$

17. $\frac{(4a)^3}{(-2a^2)^2}$

18. $\frac{(-4a^2)^2}{(2a)^2}$

19. $\frac{-60a^3b^7}{80ab^3}$

20. $\frac{15a^2b^5}{-90a^3b^3}$

21. $\frac{-6x^2y^5}{33y^3z^3}$

22. $\frac{-16x^7y^5}{-64x^3y^2}$

23. $\frac{-a^9b^3c^8}{a^6bc^7}$

24. $\frac{(-4x^2y)^2}{-12xy^4}$

4. ____________
5. ____________
6. ____________
7. ____________
8. ____________
9. ____________
10. ____________
11. ____________
12. ____________
13. ____________
14. ____________
15. ____________
16. ____________
17. ____________
18. ____________
19. ____________
20. ____________
21. ____________
22. ____________
23. ____________
24. ____________

Name Score

Write in scientific notation.

1. 0.0000356 **2.** 320,000 **3.** 0.00000037

4. 0.0000000051 **5.** 60,000,000,000 **6.** 46,000,000

1. ____________
2. ____________
3. ____________
4. ____________
5. ____________
6. ____________

Write in decimal notation.

7. 2.14×10^{-6} **8.** 5.3×10^{-9} **9.** 6.7×10^{9}

10. 5.23×10^{6} **11.** 4.5×10^{-3} **12.** 5.24×10^{7}

7. ____________
8. ____________
9. ____________
10. ____________
11. ____________
12. ____________

Simplify.

13. $(4 \times 10^{-10})(7 \times 10^{7})$ **14.** $(7.8 \times 10^{-4})(2.1 \times 10^{-5})$

15. $\dfrac{8 \times 10^{-2}}{2 \times 10^{7}}$ **16.** $\dfrac{2.8 \times 10^{5}}{4 \times 10^{-3}}$ **17.** $\dfrac{0.00072}{40{,}000{,}000}$

13. ____________
14. ____________
15. ____________
16. ____________
17. ____________

Solve. Write the answer in scientific notation.

18. A high-speed centrifuge makes 8×10^{7} revolutions each minute. Find the time in seconds for the centrifuge to make 1 revolution.

19. A space vehicle travels at an average velocity of 6.0×10^{5} miles per day. How far does it travel each week?

18. ____________
19. ____________

Name Score

Simplify.

1. $\frac{4x+4}{4}$

2. $\frac{7y+7}{7}$

3. $\frac{8a-12}{4}$

4. $\frac{20a-30}{10}$

5. $\frac{12b-30}{6}$

6. $\frac{2a^2+3a}{a}$

7. $\frac{5y^2+3y}{y}$

8. $\frac{3b^3-2b}{b}$

9. $\frac{20y^2-10y}{10y}$

10. $\frac{x^3+2x^2-4x}{x}$

11. $\frac{x^4-2x^3-x^2}{x^2}$

12. $\frac{a^6-6a^5-4a^4}{a^3}$

13. $\frac{4x^2y^2-12xy}{4xy}$

14. $\frac{10y^6-25y^4}{-5y^4}$

15. $\frac{8x^4-6x^2}{-2x^2}$

16. $\frac{3y^4-12y^3}{-3y^3}$

17. $\frac{6x^2-9x^4}{-3x^2}$

18. $\frac{4x^2+x-5}{x}$

19. $\frac{6y^2+7y-5}{y}$

20. $\frac{-20x^2+5x-4}{x}$

21. $\frac{8a^2b-10ab+12ab^2}{2ab}$

22. $\frac{20a^2b+10ab-40ab^2}{10ab}$

23. $\frac{5x^2y+6xy-2xy^2}{xy}$

24. $\frac{6ab^2-24ab+18a^2b^2}{6ab}$

1. ________
2. ________
3. ________
4. ________
5. ________
6. ________
7. ________
8. ________
9. ________
10. ________
11. ________
12. ________
13. ________
14. ________
15. ________
16. ________
17. ________
18. ________
19. ________
20. ________
21. ________
22. ________
23. ________
24. ________

Name Score

Simplify.

1. $(x^2+6x+9)\div(x+3)$

2. $(x^2+8x+16)\div(x+4)$

3. $(x^2-12x+36)\div(x-6)$

4. $(a^2-14a+49)\div(a-7)$

5. $(x^2-2x-8)\div(x+2)$

6. $(y^2-y-30)\div(y-6)$

7. $(x^2-3x-18)\div(x-6)$

8. $(y^2+5y-14)\div(x+7)$

9. $(2x^2+5x+3)\div(x+1)$

10. $(2y^2-11y+14)\div(y-2)$

11. $(2y^2+y-15)\div(y+3)$

12. $(2y^2+5y-12)\div(y+4)$

13. $(3x^2-27)\div(3x-9)$

14. $(2y^2+11y-21)\div(2y-3)$

15. $(x^2-1)\div(x+1)$

16. $(x^2+9)\div(x+3)$

17. $(4x^2-10x)\div(2x-3)$

18. $(x^2+9)\div(x+3)$

19. $(a^2+3a+10)\div(a+5)$

20. $(b^2-7b-8)\div(b-4)$

21. $(10y^2-29y-21)\div(2y-7)$

22. $(12a^2-6a-6)\div(2a+3)$

23. $(6a^2+11a-10)\div(3a-2)$

24. $(15x^2+14x-8)\div(5x-2)$

1. ______________
2. ______________
3. ______________
4. ______________
5. ______________
6. ______________
7. ______________
8. ______________
9. ______________
10. ______________
11. ______________
12. ______________
13. ______________
14. ______________
15. ______________
16. ______________
17. ______________
18. ______________
19. ______________
20. ______________
21. ______________
22. ______________
23. ______________
24. ______________

Name Score

Factor.

1. $4a+4$
2. $6c-6$
3. $8-4a^2$
4. $9+21x^2$
5. $3x+9$
6. $10a^2+15$
7. $24a-8$
8. $24x+12$
9. $9x-6$
10. $12xy-16y$
11. $6b^2-5b^3$
12. $20x^3-24x^2$
13. $12a^5-36a^2$
14. $24a^3b^4-18a^2b^2$
15. $4a^5b+6ab^4$
16. $25x^2y^2-15x^3y$
17. $3x^2y-5xy^2$
18. $8a^3b^2-12a^2b^3$
19. x^3-5x^2+7x
20. $4x^2-16x+20$
21. $6y^2-9y+12$
22. $b^4-3b^3+7b^2$
23. $4x^2-8x^3+12x^4$
24. $12y^2-16y+48$
25. $5y^4+10y^3-35y$
26. $4x^4+12x^3-28x^2$
27. $45a^4b^2-75a^3b+30ab^4$

1. ______
2. ______
3. ______
4. ______
5. ______
6. ______
7. ______
8. ______
9. ______
10. ______
11. ______
12. ______
13. ______
14. ______
15. ______
16. ______
17. ______
18. ______
19. ______
20. ______
21. ______
22. ______
23. ______
24. ______
25. ______
26. ______
27. ______

Name Score

Factor.

1. $x(a+b)+3(a+b)$
2. $a(x-y)+5(x-y)$
3. $x(b-1)-y(b-1)$
4. $y(a-1)-(a-1)$
5. $a(y+3)-(y+3)$
6. $3x(y-7)-(y-7)$
7. $2x(x-5)-(x-5)$
8. $5y(x-3)-(3-x)$
9. $x(a-2b)+y(a-2b)$
10. $a(x-9)-(x-9)$
11. $b(a-5)-2(a-5)$
12. $c(x-3y)+d(x-3y)$
13. $3x(a+1)+4(a+1)$
14. $ax-4x+4y-ay$
15. $ax-3bx+3by-ay$
16. $ax-9x-9+a$
17. $ax-7a+21-3x$
18. $2ax-8x-4y+ay$
19. $am-9m-9n+an$
20. $3mn-6m-2+n$
21. $2ab-2a-bc+c$
22. $ax-2x-3ay+6y$
23. $2a-2b+bc-ac$
24. $mn-m+2-2n$
25. $cx-7x-7y+cy$
26. $ab-ac-3c+3b$
27. $5x-5y-yz+xz$

1. ____________
2. ____________
3. ____________
4. ____________
5. ____________
6. ____________
7. ____________
8. ____________
9. ____________
10. ____________
11. ____________
12. ____________
13. ____________
14. ____________
15. ____________
16. ____________
17. ____________
18. ____________
19. ____________
20. ____________
21. ____________
22. ____________
23. ____________
24. ____________
25. ____________
26. ____________
27. ____________

Name Score

Factor.

1. $a^2-2a-35$

2. a^2-4a+3

3. $a^2+3a-10$

4. $b^2-7b+10$

5. $b^2+8b+15$

6. $b^2+6b-16$

7. y^2-y-30

8. $y^2+5y-66$

9. $x^2-4x-60$

10. $y^2-7y+10$

11. $y^2-9y+18$

12. $x^2-12x+36$

13. $a^2+3b-28$

14. $x^2+10x+16$

15. $b^2-11b-180$

16. $x^2+10x+25$

17. $b^2+10b+16$

18. $x^2-7x-60$

19. $x^2+10x-56$

20. $x^2-8x-128$

21. $x^2-4x-77$

22. $b^2-20b+84$

23. $b^2-21b+108$

24. $b^2-27b+180$

25. $a^2+16a+63$

26. $x^2-19x+60$

27. $x^2-25x+84$

1. ____________
2. ____________
3. ____________
4. ____________
5. ____________
6. ____________
7. ____________
8. ____________
9. ____________
10. ____________
11. ____________
12. ____________
13. ____________
14. ____________
15. ____________
16. ____________
17. ____________
18. ____________
19. ____________
20. ____________
21. ____________
22. ____________
23. ____________
24. ____________
25. ____________
26. ____________
27. ____________

Name Score

Factor.

1. $2x^2+8x+6$
2. $3x^2+15x+12$
3. $3a^2+3a-36$
4. $4x^2-8x-12$
5. $ab^2+5ab-14a$
6. $xy^2-7xy+10x$
7. $xy^2+11xy+28x$
8. x^3-8x^2+12x
9. $3y^3-18y^2+15y$
10. $4x^2+28x-32$
11. $2x^2-2x-12$
12. $3z^2-21z-24$
13. z^3+14z^2+45z
14. c^3+17c^2+30c
15. $x^2-9xy+14y^2$
16. $4y^2-16-128$
17. $3x^2+3x-18$
18. $2x^2-14x-60$
19. $3y^2-24y+45$
20. $5y^2-50y+45$
21. $2x^3+2x^2-12x$
22. $y^2+7xy-8x^2$
23. x^3+3x^2-40x
24. $10x^3+60x^2+80x$
25. x^4+2x^2-63
26. $2x^2+14x-240$
27. $a^4+a^3-72a^2$

1. __________
2. __________
3. __________
4. __________
5. __________
6. __________
7. __________
8. __________
9. __________
10. __________
11. __________
12. __________
13. __________
14. __________
15. __________
16. __________
17. __________
18. __________
19. __________
20. __________
21. __________
22. __________
23. __________
24. __________
25. __________
26. __________
27. __________

Name Score

Factor.

1. $2x^2 - 5x + 2$

2. $3x^2 - 2x - 1$

3. $3x^2 + x - 2$

4. $2b^2 - 13b + 6$

5. $3a^2 - 7a + 2$

6. $4x^2 + 4x - 3$

7. $5a^2 + 13a - 6$

8. $6y^2 + 5y - 6$

9. $6x^2 + x - 5$

10. $5x^2 - 3x - 2$

11. $7y^2 + 8y + 1$

12. $14x^2 + 9x + 1$

13. $7y^2 + 18y + 8$

14. $9a^2 - 3a - 2$

15. $3a^2 - 5a - 12$

16. $3x^2 - 10x - 8$

17. $6x^2 - 5x - 6$

18. $4y^2 + 25y + 25$

19. $7x^2 + 20x - 3$

20. $5x^2 + 2x - 7$

21. $10x^2 - 11x - 6$

22. $8x^2 - 26x + 15$

23. $12x^2 - 7x - 10$

24. $8x^2 - 2x - 15$

25. $10x^2 - 21x - 10$

26. $15x^2 - 26x + 8$

27. $12y^2 - 16y - 3$

1. ______
2. ______
3. ______
4. ______
5. ______
6. ______
7. ______
8. ______
9. ______
10. ______
11. ______
12. ______
13. ______
14. ______
15. ______
16. ______
17. ______
18. ______
19. ______
20. ______
21. ______
22. ______
23. ______
24. ______
25. ______
26. ______
27. ______

Name Score

Factor.

1. $3x^2-6x+3$

2. $8x^2+22x-6$

3. $6y^2-20y+14$

4. $15y^2-5y-10$

5. $2a^2b+8ab-24b$

6. $4x^2-25xy+6y^2$

7. $x^2-11xy+30y^2$

8. $2x^3-7x^2-4x$

9. $15+8x+x^2$

10. $21+4x-x^2$

11. $32+4x-x^2$

12. $4y^2-4y-3$

13. $12y^2+12y-9$

14. $12x^2-19x+4$

15. $5x^2-35x-40$

16. $5x^3-6x^2+x$

17. $3x^2-11x-20$

18. $12y^2-12y-9$

19. $8x^2+8x-6$

20. $10a^3-15a^2-25a$

21. $6x^3+17x^2+5x$

22. $6x^3-26x^2+8x$

23. $10b^2+14b-12$

24. $22z-9z^2-z^3$

25. $6x^3-32x^2+10x$

26. $19+18x-x^2$

27. $13x^2+49x-12$

1. ________
2. ________
3. ________
4. ________
5. ________
6. ________
7. ________
8. ________
9. ________
10. ________
11. ________
12. ________
13. ________
14. ________
15. ________
16. ________
17. ________
18. ________
19. ________
20. ________
21. ________
22. ________
23. ________
24. ________
25. ________
26. ________
27. ________

Name Score

Factor.

1. $x^2 - 16$
2. $x^2 - 64$
3. $9x^2 - 1$
4. $16x^2 - 25$
5. $x^8 - 100$
6. $81x^2 - 1$
7. $y^4 - 121$
8. $x^2 + 25$
9. $x^4 - y^8$
10. $b^2 - 144c^2$
11. $x^6 - 81$
12. $x^2y^2 - 1$
13. $x^2 - 400$
14. $36a^2 - 1$
15. $49x^2 - 4$
16. $x^2 + 6x + 9$
17. $a^2 - 2a + 4$
18. $x^2 + 16x + 64$
19. $25 - 10a + a^2$
20. $36x^2 - 12x + 1$
21. $100x^2 - 20x + 1$
22. $4x^2 + 36x + 81$
23. $4a^2 + 44a + 121$
24. $49a^2 + 70a + 25$
25. $4x^2 - 20xy + 25y^2$
26. $4x^2 + 12xy + 9y^2$
27. $9x^2 - 30xy + 25y^2$

1. ______
2. ______
3. ______
4. ______
5. ______
6. ______
7. ______
8. ______
9. ______
10. ______
11. ______
12. ______
13. ______
14. ______
15. ______
16. ______
17. ______
18. ______
19. ______
20. ______
21. ______
22. ______
23. ______
24. ______
25. ______
26. ______
27. ______

Name Score

Factor.

1. $3x^2 - 12$
2. $2x^2 - 50$
3. $x^3 + 2x^2 + x$
4. $y^3 + 8y^2 + 16y$
5. $x^4 + x^3 - 6x^2$
6. $a^4 - 3a^3 + 40a^2$
7. $3b^2 + 30b + 63$
8. $4y^2 - 32y + 28$
9. $2a^2 - 18a - 44$
10. $x^3 - 8x^2 - 20x$
11. $b^3 - 5b^2 - 6b$
12. $5a^3 - 30a^2 + 45a$
13. $4x^2 - 6x + 2$
14. $21x^4 - 11x^3 + 5x^2$
15. $x^4 - 16x^2$
16. $15x^3 - 18x^2 + 3x$
17. $9a^3 + 39a^2 + 12a$
18. $3xy^2 + 11xy - 20x$
19. $24 + 6x - 3x^2$
20. $a^2b^2 + 7ab^2 - 8b^2$
21. $4x^2y + 12xy + 8y$
22. $18a^3 + 54a^2 + 36a$
23. $x^2y^2 - y^2$
24. $5x^2 - 45y^2$
25. $x^4 - 9x^2$
26. $a(3x - 3) + b(3x - 3)$
27. $y^2(a + b) - 4(a + b)$

1. ____________
2. ____________
3. ____________
4. ____________
5. ____________
6. ____________
7. ____________
8. ____________
9. ____________
10. ____________
11. ____________
12. ____________
13. ____________
14. ____________
15. ____________
16. ____________
17. ____________
18. ____________
19. ____________
20. ____________
21. ____________
22. ____________
23. ____________
24. ____________
25. ____________
26. ____________
27. ____________

Name Score

Solve.

1. $(y+1)(y+2)=0$

2. $(y-4)(y-6)=0$

3. $(z-6)(z-1)=0$

4. $(x+7)(x-5)=0$

5. $x(x-8)=0$

6. $x(x+1)=0$

7. $a(a-4)=0$

8. $y(3y+2)=0$

9. $t(2t-5)=0$

10. $3a(2a-1)=0$

11. $2b(4b+3)=0$

12. $(b+1)(b-4)=0$

13. $(b-7)(b+4)=0$

14. $x^2-4x-21=0$

15. $x^2+6x-16=0$

16. $x^2-5x=6$

17. $x^2-5x=14$

18. $2a^2-a=3$

19. $4t^2-13=-3$

20. $5a^2+13a=6$

21. $2x^2+5x=-2$

22. $x(x+10)=11$

23. $y(y-9)=-18$

24. $x(x+5)=50$

25. $(2x+3)(x-1)=25$

26. $(z+1)(z-9)=39$

27. $(y+3)(2y-1)=9$

1. ____________
2. ____________
3. ____________
4. ____________
5. ____________
6. ____________
7. ____________
8. ____________
9. ____________
10. ____________
11. ____________
12. ____________
13. ____________
14. ____________
15. ____________
16. ____________
17. ____________
18. ____________
19. ____________
20. ____________
21. ____________
22. ____________
23. ____________
24. ____________
25. ____________
26. ____________
27. ____________

Name Score

Solve.

1. The square of a positive number is four more than three times the positive number. Find the number.

2. The square of a negative number is forty-eight more than twice the negative number. Find the number.

3. The sum of two numbers is seven. The sum of the squares of the two numbers is twenty-nine. Find the two numbers.

4. The sum of two numbers is ten. The sum of the squares of the two numbers is fifty-two. Find the two numbers.

5. The sum of two numbers is twelve. The product of the two numbers is twenty-seven. Find the two numbers.

6. The square of the sum of a number and six is eighty-one. Find the number.

7. The sum of the squares of two consecutive positive integers is sixty-one. Find the two integers.

8. The product of two consecutive even positive integers is one hundred twenty. Find the integers.

9. The length of the base of a triangle is three times the height. The area of the triangle is 96 ft^2. Find the base and height of the triangle.

10. The length of a rectangle is four times the width. The area is 144 in^2. Find the length and width.

11. The width of a rectangle is 7 ft less than the length. The area of the rectangle is 120 ft^2. Find the length and width of the rectangle.

12. The length of each side of a square is extended by 3 in. The area of the resulting square is 121 in^2. Find the length of a side of the original square.

1. ____________

2. ____________

3. ____________

4. ____________

5. ____________

6. ____________

7. ____________

8. ____________

9. ____________

10. ____________

11. ____________

12. ____________

Name Score

Simplify.

1. $\dfrac{8x^2}{12x^4}$

2. $\dfrac{24x^2y}{32xy^3}$

3. $\dfrac{(x+1)^3}{(x+1)^5}$

4. $\dfrac{(3x-2)^6}{(3x-2)^4}$

5. $\dfrac{2n-3}{3-2n}$

6. $\dfrac{15y^2(y+3)}{20y^3(y+3)}$

7. $\dfrac{15y(y+1)}{45y^2(y+1)}$

8. $\dfrac{24x^3(4-x)}{18x(4-x)}$

9. $\dfrac{6x^3(2-x)}{12x(2-x)}$

10. $\dfrac{a^2+6a}{ac+6c}$

11. $\dfrac{x^2+4x}{3x+12}$

12. $\dfrac{6-8x}{4x^3-3x}$

13. $\dfrac{x^2+2x-15}{x^2-10x+21}$

14. $\dfrac{x^2-2x-3}{x^2-5x-6}$

15. $\dfrac{x^2-7x+12}{x^2-x-12}$

16. $\dfrac{x^2-3x-10}{x^2-6x+5}$

17. $\dfrac{x^2-8x+16}{x^2-9x+20}$

18. $\dfrac{x^2-7x+10}{x^2-4x-5}$

19. $\dfrac{x^2-2x-24}{16-x^2}$

20. $\dfrac{9-y^2}{y^2+2y-15}$

21. $\dfrac{3x^3+18x^2-21x}{x^3+5x^2-14x}$

22. $\dfrac{2x^2-18x}{4x^3-32x^2-36x}$

23. $\dfrac{2x^2-7x+3}{2x^2-9x-5}$

24. $\dfrac{x^2+2x-24}{12-4x-x^2}$

25. $\dfrac{2x^2-8}{x^2+4x-12}$

26. $\dfrac{2x^2-9x+7}{x^2+2x-3}$

27. $\dfrac{8x^2-14x+6}{4x-3}$

1. ______________
2. ______________
3. ______________
4. ______________
5. ______________
6. ______________
7. ______________
8. ______________
9. ______________
10. ______________
11. ______________
12. ______________
13. ______________
14. ______________
15. ______________
16. ______________
17. ______________
18. ______________
19. ______________
20. ______________
21. ______________
22. ______________
23. ______________
24. ______________
25. ______________
26. ______________
27. ______________

Name Score

Multiply.

1. $\frac{9xy}{35a^3b^2}\cdot\frac{14a^2b^3}{45x^3y^2}$

2. $\frac{8xy^3}{75a^2b^2}\cdot\frac{15a^3b^2}{32x^2y^2}$

3. $\frac{9x-18}{4x+16}\cdot\frac{20x+80}{12x-24}$

4. $\frac{6x-9}{12x+6}\cdot\frac{36x+18}{24x-36}$

5. $\frac{x^2-16}{x^2-4x}\cdot\frac{x^2+x}{x^2+8x+16}$

6. $\frac{x^2-2x+1}{x^2+x-2}\cdot\frac{x^2+8x}{x^2+7x-8}$

7. $\frac{x^2-3x-10}{x^2-5x}\cdot\frac{x+2}{x^2-4}$

8. $\frac{x^2-x-20}{x^2y^3}\cdot\frac{x^3y^2}{x^2-10x+25}$

9. $\frac{3x^2-7x}{3xy+y}\cdot\frac{3xy^2+y^2}{6x^2-14x}$

10. $\frac{xy^2}{x^2+12x+32}\cdot\frac{x^2+3x-4}{x^3y^2}$

11. $\frac{x^2+5x-14}{x^2+6x-16}\cdot\frac{x^2-64}{x^2-x-56}$

12. $\frac{x^2-3x-10}{x^2+6x+8}\cdot\frac{x^2-x-20}{x^2-25}$

13. $\frac{x^2-7x-18}{x^2-81}\cdot\frac{x^2-11x+18}{x^2-4}$

14. $\frac{x^2+2x-15}{x^2+12x+35}\cdot\frac{x^2+16x+63}{x^2-x-6}$

15. $\frac{x^2+5x-14}{x^2-x-6}\cdot\frac{x^2+x-2}{x^2-3x+2}$

16. $\frac{x^2-3x-54}{x^2-8x-20}\cdot\frac{x^2-9x-10}{x^2-8x-9}$

17. $\frac{x^2-4x-32}{x^2-8x}\cdot\frac{x^3+2x^2+x}{x^2+5x+4}$

18. $\frac{2x^2+3x-20}{2x^2+7x-30}\cdot\frac{4x^2-12x+9}{2x^2+5x-12}$

19. $\frac{3x^2-10x+3}{2x^2-5x-3}\cdot\frac{2x^2-3x-2}{2x^2-x-6}$

20. $\frac{x^2+4x+3}{6x^2+17x+12}\cdot\frac{3x^2+x-4}{x^2-1}$

1. ______________
2. ______________
3. ______________
4. ______________
5. ______________
6. ______________
7. ______________
8. ______________
9. ______________
10. ______________
11. ______________
12. ______________
13. ______________
14. ______________
15. ______________
16. ______________
17. ______________
18. ______________
19. ______________
20. ______________

Name Score

Divide.

1. $\dfrac{6x^2y^3}{25a^2b^2} \div \dfrac{8xy}{15ab}$

2. $\dfrac{18a^2b^4}{25a^3b^2} \div \dfrac{27a^3b^2}{100a^4b^2}$

3. $\dfrac{5x-10}{6x-24} \div \dfrac{15x-30}{9x-36}$

4. $\dfrac{18x+9}{24x-36} \div \dfrac{12x+6}{20x-30}$

5. $\dfrac{x^2+2x}{x^2-4} \div \dfrac{x+2}{x-2}$

6. $\dfrac{x^2-x-6}{x^2+2x+1} \div \dfrac{x+2}{x+1}$

7. $\dfrac{x^2+3x+2}{xy^2} \div \dfrac{x^2+4x+4}{x^2y^2}$

8. $\dfrac{x^2-4}{x+2} \div \dfrac{x-2}{x+3}$

9. $\dfrac{x^2-25}{x^3y^4} \div \dfrac{x^2-10x+25}{x^4y^2}$

10. $\dfrac{6ax-9a}{y^2} \div \dfrac{2x-3}{y^4}$

11. $\dfrac{x^2-5x-14}{x^2-9x+14} \div \dfrac{x^2+6x+8}{x^2+2x-8}$

12. $\dfrac{x^2+x-6}{x^2-9} \div \dfrac{x^2-9x+14}{x^2-8x+15}$

13. $\dfrac{x^2-3x+2}{x^2-7x+10} \div \dfrac{x^2-1}{x^2-4x-5}$

14. $\dfrac{x^2-x-30}{x^2+6x+5} \div \dfrac{x^2-11x+30}{x^2-6x+5}$

15. $\dfrac{x^2-2x-63}{x^2-4x-45} \div \dfrac{x^2+3x-28}{x^2+9x+20}$

16. $\dfrac{x^2-11x+28}{x^2-13x+42} \div \dfrac{x^2-9x+20}{x^2+x-30}$

17. $\dfrac{x^2-6x-40}{x^2-7x-30} \div \dfrac{2x^2+7x+3}{2x^2+5x+2}$

18. $\dfrac{3x^2+17x+10}{3x^2-22x-16} \div \dfrac{x^2-8x-48}{x^2-4x-32}$

19. $\dfrac{2x^2+5x-3}{x^2-2x-15} \div \dfrac{2x^2-5x+2}{x-5}$

20. $\dfrac{3y^2-14y+8}{2y^2-7y-4} \div \dfrac{3y^2-16y+5}{2y^2-9y-5}$

1. ______
2. ______
3. ______
4. ______
5. ______
6. ______
7. ______
8. ______
9. ______
10. ______
11. ______
12. ______
13. ______
14. ______
15. ______
16. ______
17. ______
18. ______
19. ______
20. ______

Name Score

Find the LCM of the polynomials.

1. $6x^2y$
   $9xy^2$

2. $10ab^3$
   $30ab^2$

3. $12x^2y^3$
   $30x^3y$

4. $15x^3y$
   $20xy^2$

5. $10x^3$
   $5x^2+10x$

6. $8x^3$
   $12x+8$

7. $3xy^2$
   $4x^2+12x$

8. $8x(x+1)^2$
   $18(x+1)$

9. $12x^3(x-1)$
   $18x^2(x-1)^2$

10. $2x+6$
    $2x^2+8x+6$

11. $3x-12$
    $3x^2-15x+12$

12. $(x-2)(x+3)$
    $(x-2)(x+4)$

13. $(2x-3)(x+2)$
    $(2x+3)(x+2)$

14. $(3x+1)^2$
    $(3x+1)(x-4)$

15. $(x+2)$
    $(x-3)$
    $(x+2)(x-3)$

16. x^2+x-20
    $x^2+2x-15$

17. $x^2-3x-18$
    $x^2-2x-24$

18. x^2-6x+5
    x^2+6x-7

19. $x^2-9x+18$
    $x^2-8x+12$

20. x^2+9x+8
    x^2-64

21. x^2-x-42
    $x^2-12x+35$

22. $2x^2+x-6$
    $2x^2-7x+6$

23. $2x^2-15x+7$
    $2x^2+9x-5$

24. $24+2x-x^2$
    $x-6$
    $x+4$

25. $14+5x-x^2$
    $x-7$
    $x+2$

26. $x^2+2x-15$
    $3-x$
    $x+5$

27. $x^2-8x+12$
    $2-x$
    $x-6$

1. ____________
2. ____________
3. ____________
4. ____________
5. ____________
6. ____________
7. ____________
8. ____________
9. ____________
10. ____________
11. ____________
12. ____________
13. ____________
14. ____________
15. ____________
16. ____________
17. ____________
18. ____________
19. ____________
20. ____________
21. ____________
22. ____________
23. ____________
24. ____________
25. ____________
26. ____________
27. ____________

Name Score

Write each fraction in terms of the LCM of the denominators.

1. $\frac{2}{x^2}, \frac{5}{x}$

2. $\frac{4}{a^2b}, \frac{7}{ab^2}$

3. $\frac{x}{2y}, \frac{z}{5y^2}$

4. $\frac{3}{6y^2}, \frac{5}{4x^2y}$

5. $\frac{x}{y}, \frac{5}{y(y-1)}$

6. $\frac{a}{y}, \frac{3}{y(y+2)}$

7. $\frac{5}{x-2}, \frac{7}{x^2(x-2)}$

8. $\frac{a}{y(y+6)}, \frac{a^2}{(y+6)^2}$

9. $\frac{4}{x-2}, \frac{7}{x(2-x)}$

10. $\frac{4}{(x-1)^2}, \frac{2}{x-1}$

11. $\frac{6}{6-y}, \frac{3}{(y-6)^2}$

12. $\frac{a}{x(x-3)}, \frac{a^2}{3-x}$

13. $\frac{2}{(x-6)^2}, \frac{3}{6-x}$

14. $\frac{5}{x^2+3x}, \frac{6}{x^2}$

15. $\frac{3}{x-2}, \frac{4}{x^2-2x}$

16. $\frac{x-1}{x+2}, \frac{x}{x-3}$

17. $\frac{x}{x-7}, \frac{x+2}{x+3}$

18. $\frac{2}{x^2-2x-3}, \frac{x}{x-3}$

19. $\frac{2x}{x-2}, \frac{3}{x^2-4}$

20. $\frac{4}{x^2-16}, \frac{x+1}{x^2-8x+16}$

21. $\frac{x-4}{2x^2+3x-9}, \frac{3}{x+3}$

22. $\frac{x-1}{3x^2+5x-2}, \frac{3}{x+2}$

23. $\frac{x}{x^2+x-2}, \frac{3x}{x^2-1}$

24. $\frac{x-2}{x^2+3x-4}, \frac{x}{x^2-4x+3}$

25. $\frac{x}{4-x^2}, \frac{x-3}{x^2-4x+4}$

26. $\frac{x}{x-1}, \frac{2x}{x+3}, \frac{2}{3-2x-x^2}$

27. $\frac{x+2}{x+4}, \frac{x+1}{x-5}, \frac{2}{20+x-x^2}$

1. ________
2. ________
3. ________
4. ________
5. ________
6. ________
7. ________
8. ________
9. ________
10. ________
11. ________
12. ________
13. ________
14. ________
15. ________
16. ________
17. ________
18. ________
19. ________
20. ________
21. ________
22. ________
23. ________
24. ________
25. ________
26. ________
27. ________

Name Score

Simplify.

1. $\frac{2}{x^2}+\frac{5}{x^2}$

2. $\frac{9}{xy}-\frac{3}{xy}$

3. $\frac{2}{x+3}-\frac{7}{x+3}$

4. $\frac{x}{x-5}-\frac{3}{x-5}$

5. $\frac{4x}{3x+2}+\frac{6x}{3x+2}$

6. $\frac{5y}{5y-1}-\frac{8y}{5y-1}$

7. $\frac{3x+2}{x-2}+\frac{x+4}{x-2}$

8. $\frac{5x+4}{2x-1}+\frac{3x-9}{2x-1}$

9. $\frac{4x-3}{x+8}-\frac{2x-19}{x+8}$

10. $\frac{7x+4}{x-9}-\frac{4x+31}{x-9}$

11. $\frac{x-6}{3x+5}-\frac{5x-4}{3x+5}$

12. $\frac{3n}{2n+9}-\frac{6n-5}{2n+9}$

13. $\frac{x}{x^2-3x-10}+\frac{2}{x^2-3x-10}$

14. $\frac{2x}{x^2-x-12}-\frac{8}{x^2-x-12}$

15. $\frac{3x+5}{x^2-6x+8}-\frac{2x+9}{x^2-6x+8}$

16. $\frac{2x-7}{x^2+4x-12}-\frac{x-13}{x^2+4x-12}$

17. $\frac{3y-5}{2y^2+9y-5}-\frac{y-4}{2y^2+9y-5}$

18. $\frac{4x-3}{2x^2+x-3}+\frac{x-2}{2x^2+x-3}$

19. $\frac{3x^2+x}{x^2-6x+5}+\frac{2x^2+3x}{x^2-6x+5}-\frac{5x^2+4}{x^2-6x+5}$

20. $\frac{2x^2-5}{x^2-x-56}-\frac{x^2+2x+10}{x^2-x-56}+\frac{x-6}{x^2-x-56}$

1. ____________
2. ____________
3. ____________
4. ____________
5. ____________
6. ____________
7. ____________
8. ____________
9. ____________
10. ____________
11. ____________
12. ____________
13. ____________
14. ____________
15. ____________
16. ____________
17. ____________
18. ____________
19. ____________
20. ____________

Name Score

Simplify.

1. $\frac{8}{x}-\frac{4}{3x}$

2. $\frac{3}{2a}-\frac{2}{3a}$

3. $\frac{2}{3x}+\frac{4}{9x}-\frac{1}{6x}$

4. $\frac{5}{y^2}+\frac{2}{3y}-\frac{3}{5y}$

5. $\frac{2x-1}{2x}+\frac{x+1}{3x}$

6. $\frac{3x-2}{8x}+\frac{x+4}{12x}$

7. $\frac{4x+7}{9x}-\frac{x-3}{4x}$

8. $\frac{x+1}{xy^2}-\frac{3x-3}{xy}$

9. $\frac{x-9}{6x^2}+\frac{x-2}{2x}$

10. $\frac{x+4}{9x}-\frac{x-4}{12x^2}$

11. $\frac{4}{x-1}-\frac{3}{x+2}$

12. $\frac{5}{y+6}-\frac{2}{y-3}$

13. $\frac{2x}{x-8}+\frac{8}{x-2}$

14. $\frac{3x}{2x-1}-\frac{6}{x-3}$

15. $\frac{x}{x^2-4}-\frac{2}{x-2}$

16. $\frac{4x}{x^2-2x-8}+\frac{3}{x-4}$

17. $\frac{x+2}{x^2-x-6}-\frac{3}{3-x}$

18. $\frac{2x+1}{x-2}-\frac{x-3}{x-1}+\frac{x+2}{x^2-3x+2}$

1. ________
2. ________
3. ________
4. ________
5. ________
6. ________
7. ________
8. ________
9. ________
10. ________
11. ________
12. ________
13. ________
14. ________
15. ________
16. ________
17. ________
18. ________

Name Score

Simplify.

1. $\dfrac{1-\dfrac{5}{x}}{1-\dfrac{25}{x^2}}$

2. $\dfrac{2+\dfrac{6}{x}}{1-\dfrac{36}{x^2}}$

3. $\dfrac{1-\dfrac{6}{x+3}}{2-\dfrac{10}{x+3}}$

4. $\dfrac{6-\dfrac{18}{x-5}}{2-\dfrac{6}{x-5}}$

5. $\dfrac{5-\dfrac{10}{2x+1}}{3-\dfrac{18}{2x+1}}$

6. $\dfrac{2-\dfrac{4}{x+6}}{3-\dfrac{6}{x+6}}$

7. $\dfrac{1-\dfrac{12}{x}+\dfrac{27}{x^2}}{1-\dfrac{9}{x^2}}$

8. $\dfrac{1+\dfrac{9}{x}+\dfrac{14}{x^2}}{1-\dfrac{49}{x^2}}$

9. $\dfrac{1-\dfrac{11}{x}+\dfrac{30}{x^2}}{1-\dfrac{2}{x}-\dfrac{15}{x^2}}$

10. $\dfrac{1+\dfrac{3}{x}-\dfrac{10}{x^2}}{1+\dfrac{8}{x}+\dfrac{15}{x^2}}$

11. $\dfrac{x-\dfrac{3x}{x+3}}{x+\dfrac{3x}{x+3}}$

12. $\dfrac{x+\dfrac{1}{x-1}}{1+\dfrac{1}{x-1}}$

13. $\dfrac{x+3+\dfrac{12}{x+2}}{x+2-\dfrac{3}{x+2}}$

14. $\dfrac{x+2-\dfrac{7}{x-4}}{x+3-\dfrac{8}{x-4}}$

15. $\dfrac{\dfrac{2}{x}-\dfrac{1}{x+1}}{\dfrac{6}{x}+\dfrac{1}{x+1}}$

16. $\dfrac{\dfrac{4}{2x+1}-\dfrac{2}{x}}{\dfrac{6}{x}-\dfrac{3}{2x+1}}$

1. ____________
2. ____________
3. ____________
4. ____________
5. ____________
6. ____________
7. ____________
8. ____________
9. ____________
10. ____________
11. ____________
12. ____________
13. ____________
14. ____________
15. ____________
16. ____________

Name Score

Solve.

1. $\frac{2x}{5}-\frac{4}{3}=-\frac{8}{15}$

2. $\frac{x}{2}-\frac{2}{3}=\frac{5}{6}$

3. $\frac{3x-4}{6}+\frac{1}{2}=\frac{x}{4}+\frac{7}{12}$

4. $\frac{2x+3}{10}-\frac{1}{2}=\frac{4x+2}{10}-\frac{3}{5}$

5. $\frac{5}{2a-1}=3$

6. $\frac{20}{2x-3}=4$

7. $3+\frac{9}{x}=6$

8. $2+\frac{7}{n}=3$

9. $\frac{2}{x-1}=\frac{3}{x}$

10. $\frac{4}{x+2}=\frac{3}{x+1}$

11. $\frac{3}{2x-7}=\frac{2}{3x+2}$

12. $\frac{4}{4x-1}=\frac{-2}{1-3x}$

13. $\frac{x}{x-2}=3-\frac{2}{x-2}$

14. $\frac{6}{y+3}=\frac{3}{y-1}$

15. $\frac{3x}{x-2}-3=\frac{x}{x-2}$

16. $\frac{n}{n+8}=\frac{1}{n+3}$

17. $\frac{x}{x-1}=\frac{6}{x+1}$

18. $\frac{x}{x+12}=\frac{2}{x+4}$

19. $\frac{6}{y}=\frac{1}{y-1}+2$

20. $\frac{2x}{3x-4}=\frac{x-3}{x}$

1. ____________

2. ____________

3. ____________

4. ____________

5. ____________

6. ____________

7. ____________

8. ____________

9. ____________

10. ____________

11. ____________

12. ____________

13. ____________

14. ____________

15. ____________

16. ____________

17. ____________

18. ____________

19. ____________

20. ____________

Name Score

Solve.

1. $\frac{x}{18}=\frac{5}{6}$

2. $\frac{12}{x}=\frac{2}{5}$

3. $\frac{5}{8}=\frac{x}{40}$

4. $\frac{x}{20}=\frac{4}{5}$

5. $\frac{8}{x}=\frac{2}{3}$

6. $\frac{6}{7}=\frac{x}{28}$

7. $\frac{15}{8}=\frac{75}{x}$

8. $\frac{x+2}{10}=\frac{4}{5}$

9. $\frac{7}{8}=\frac{x-3}{16}$

10. $\frac{12}{7}=\frac{48}{x}$

11. $\frac{5}{6}=\frac{x-2}{18}$

12. $\frac{15}{x+6}=\frac{5}{8}$

13. $\frac{3}{10}=\frac{18}{x-4}$

14. $\frac{3}{x}=\frac{12}{x+3}$

15. $\frac{16}{x-3}=\frac{2}{3}$

16. $\frac{4}{9}=\frac{16}{x+2}$

17. $\frac{2}{x}=\frac{6}{x-1}$

18. $\frac{x+2}{5}=\frac{x}{20}$

19. $\frac{x-5}{2}=\frac{x}{7}$

20. $\frac{20}{x-3}=\frac{5}{x}$

21. $\frac{x+1}{3}=\frac{x}{9}$

22. $\frac{x-7}{4}=\frac{x}{6}$

23. $\frac{3}{x-2}=\frac{12}{2x-1}$

24. $\frac{15}{x+3}=\frac{3}{x-1}$

25. $\frac{3x}{5}=\frac{x-3}{10}$

26. $\frac{4}{x-3}=\frac{16}{3x+1}$

27. $\frac{5x}{6}=\frac{x-1}{12}$

1. __________
2. __________
3. __________
4. __________
5. __________
6. __________
7. __________
8. __________
9. __________
10. __________
11. __________
12. __________
13. __________
14. __________
15. __________
16. __________
17. __________
18. __________
19. __________
20. __________
21. __________
22. __________
23. __________
24. __________
25. __________
26. __________
27. __________

Name Score

Solve.

1. The monthly loan payment for a car is $31.05 for each $1000 borrowed. At this rate, find the monthly payment for a $7000 car loan.

2. An investment of $1000 earns $130 each year. At the same rate, how much additional money must be invested to earn $325 each year?

3. Eighteen ceramic tiles are required to tile a 12-square-foot area. How many feet can be tiled using 324 ceramic tiles?

4. A pre-election survey showed that 4 out of every 5 voters would vote for a new library. At this rate, how many people would be expected to vote for the library if 20,000 votes were cast?

5. A quality control inspector found 3 defective electric clocks in a shipment of 100 clocks. At this rate, how many clocks would be defective in a shipment of 3000?

6. A license fee for a car which cost $6000 was $72. At the same rate, what is the license fee for a car which costs $8000?

7. In a wildlife preserve, 20 deer are captured, tagged, and then released. Later, 50 deer are captured. Four of the 50 are found to have tags. Estimate the number of deer in the preserve.

8. A stock investment of 100 shares paid a dividend of $140. At this rate, how many additional shares are required to earn a dividend of $210?

9. Two ounces of a medication are required for a 120-pound adult. At the same rate, how many additional ounces of medication are required for a 180-pound adult?

10. A caterer estimates that 2 gal of fruit punch will serve 30 people. How much additional punch is necessary to serve 75 people?

1. ________
2. ________
3. ________
4. ________
5. ________
6. ________
7. ________
8. ________
9. ________
10. ________

Name ____________________ Score ____________________

Triangles *ABC* and *DEF* and similar. Find the indicated distance. Round to the nearest tenth.

1. Find side *AC*.

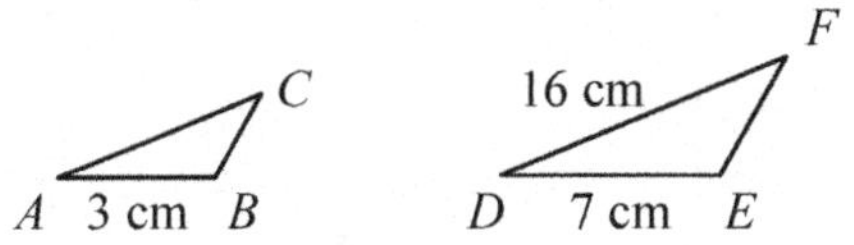

2. Find side *DE*.

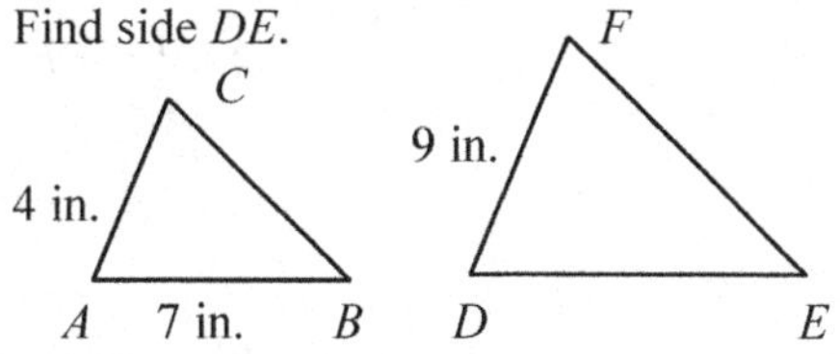

3. Find the height of triangle *ABC*.

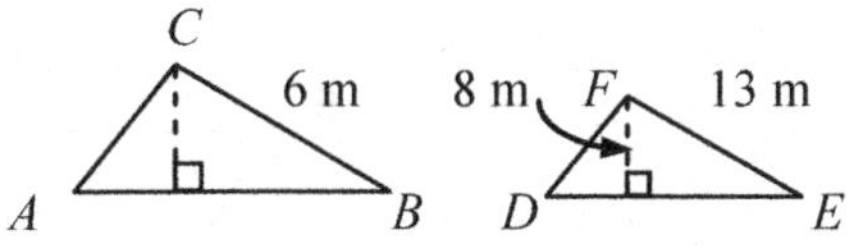

4. Find the perimeter of triangle *ABC*.

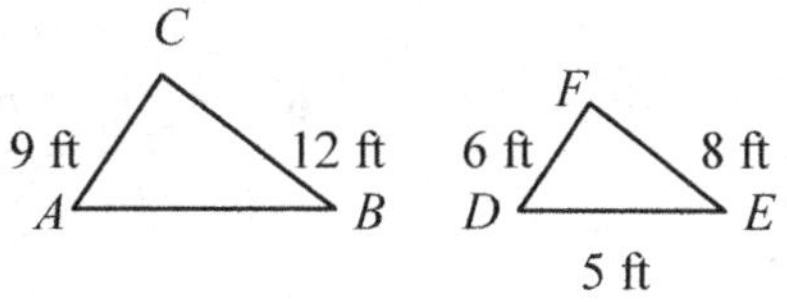

5. Find the perimeter of triangle *ABC*.

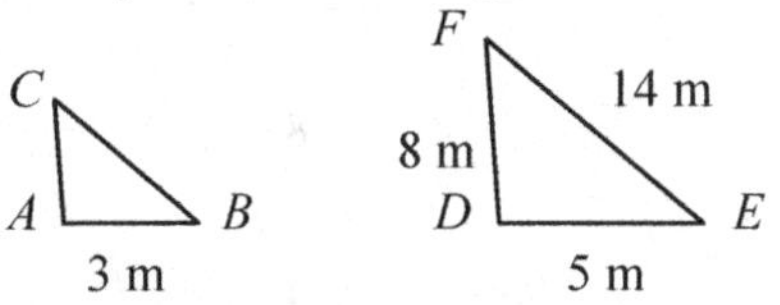

6. Find the area of triangle *ABC*.

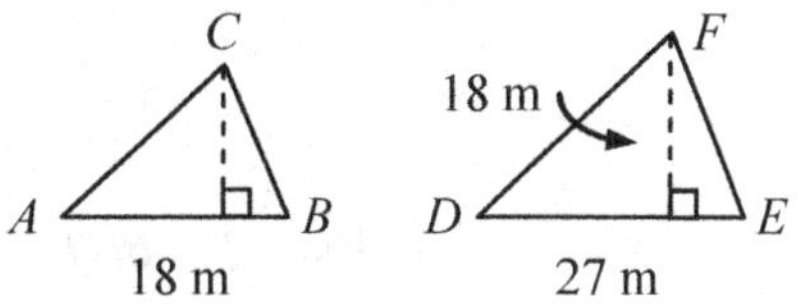

7. Find the area of triangle *DEF*.

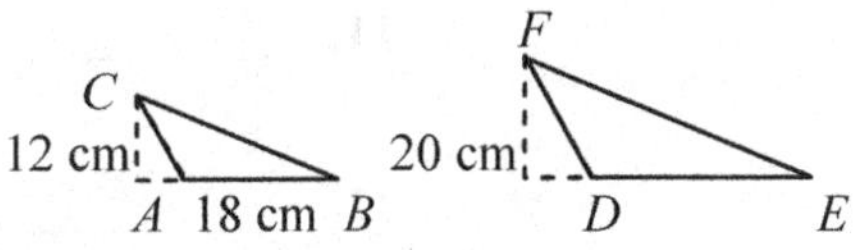

8. Given $BD \parallel AE$, *BD* measures 6 cm, *AE* measures 10 cm, and *AC* measures 12 cm. Find the length of *BC*.

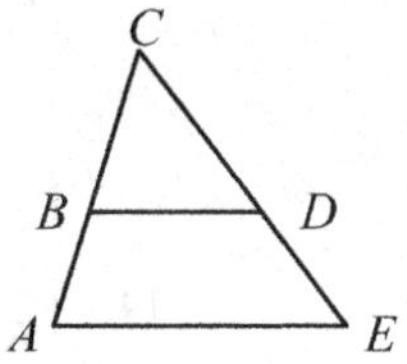

9. Given $DE \parallel AC$, *DE* measures 9 in., *AC* measures 15 in., and *AB* measures 22.5 in. Find the length of *DA*.

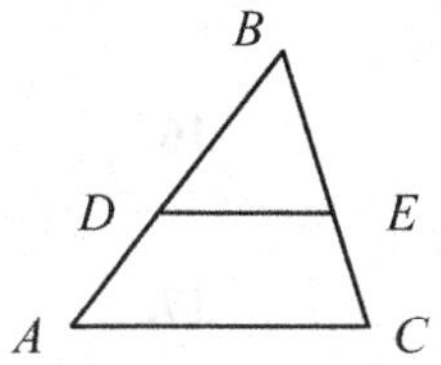

10. The sun's rays cast a shadow as shown in the diagram. Find the height of the flagpole. Write the answer in terms of feet.

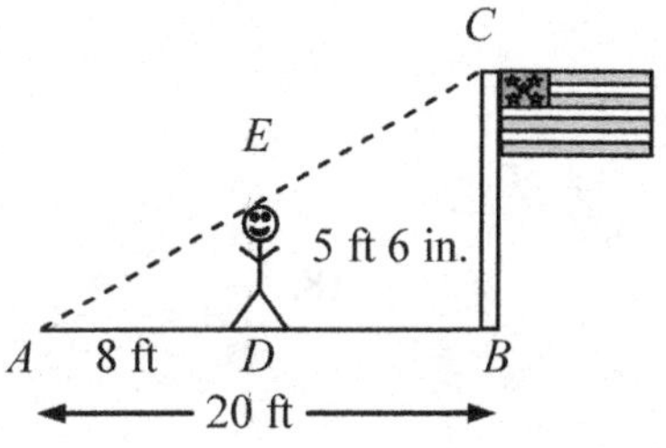

1. ____________

2. ____________

3. ____________

4. ____________

5. ____________

6. ____________

7. ____________

8. ____________

9. ____________

10. ____________

Name Score

Solve the formula for the given variable.

1. $a_n = a_1 + (n-1)d;\ n$

2. $A = \frac{1}{2}bh;\ b$

3. $x = -\frac{b}{2a};\ a$

4. $PV = nRT;\ n$

5. $F = \frac{Gm_1m_2}{r^2};\ G$

6. $\frac{P_1V_1}{T_1} = \frac{P_2V_2}{T_2};\ V_2$

7. $A = \frac{1}{2}h(b_1 + b_2);\ b_1$

8. $d = rt;\ r$

9. $I = \frac{100M}{C};\ C$

10. $E = aI(T - t);\ T$

11. $\frac{1}{t} = \frac{1}{a} + \frac{1}{b};\ a$

12. $d = \frac{a}{2}(2t - v);\ v$

13. $h = vt + \frac{1}{2}gt^2;\ g$

14. $R = \frac{C - S}{t};\ C$

15. $R = \frac{MB - T}{K};\ M$

16. $K = mr - mv;\ m$

17. $S = \pi r(r - I);\ I$

18. $A = \frac{P}{1 - c};\ c$

1. __________
2. __________
3. __________
4. __________
5. __________
6. __________
7. __________
8. __________
9. __________
10. __________
11. __________
12. __________
13. __________
14. __________
15. __________
16. __________
17. __________
18. __________

Name Score

Solve.

1. Two farmers are planting peanuts. One farmer, using an old planter and working alone, required 6 h to plant the crop. A second farmer, using a modern planter, can plant the crop in 2 h. How long would it take to plant the peanut crop with both planters working together?

2. A high speed copier can completely copy a report in 10 min. An older copier requires 15 min to do the same task. How long would it take to copy the report with both copiers operating?

3. One member of a golf course greens crew can mow a fairway in 40 min, while the second member of the crew requires 24 min to mow the same fairway. How long would it take to mow the fairway when they work together?

4. A small heating unit will heat a room 10° in 60 min. A larger heater will heat the room 10° in 40 min. How long would it take to heat the room 10° with both heaters operating?

5. With cold and hot water running, a bathtub can be filled in 8 min. The hot water faucet working alone requires 24 min to fill the tub. How long would it take the cold water faucet working alone to fill the tub?

6. An experienced bricklayer can construct a wall twice as fast as an apprentice bricklayer. Working together, the bricklayers require 8 h to build the wall. How long would it take the apprentice working along to build the wall?

7. A large and small inlet pipe work together to fill a storage tank. The larger pipe requires 8 h working alone to fill the tank. After both pipes have been operating for 3 h, the larger pipe is turned off. The small pipe requires 10 more hours to fill the storage tank. How long would it take the smaller pipe working alone to fill the tank?

8. One welder requires 20 h to make the welds in a steel frame, while a second welder requires 35 h to do the same job. The first welder worked for 8 hours and then quit. How long would it take the second welder to complete the welds?

9. Two computer printers that work at the same rate are working together to print the payroll checks for a large corporation. After working together for 3 h, one of the printers quits. The second requires 2 more hours to complete the payroll checks. Find the time it would take one printer alone to print the payroll.

10. Two roofers who work at the same rate are roofing a section of an apartment. After working together for 8 h, one roofer quits. The second roofer required 16 more hours to finish the job. How long would it take one of the roofers working alone to roof the apartment?

1. ____________

2. ____________

3. ____________

4. ____________

5. ____________

6. ____________

7. ____________

8. ____________

9. ____________

10. ____________

Name Score

Solve.

1. An express train travels 440 mi in the same amount of time that a freight train travels 280 mi. The rate of the express train is 20 mph faster than the freight train. Find the rate of each train.

2. A commuter plane flew 720 mi in the same amount of time it takes a jet to fly 1520 mi. The rate of the jet is 200 mph faster than the commuter plane. Find the rate of each plane.

1. ____________

2. ____________

3. A sales accountant traveled 75 mi by car and 1800 mi by jet plane. The rate of the jet is eight times the rate of the car. The entire trip took 6 h. Find the rate of each.

4. A single-engine plane and a corporate jet leave an airport at noon and head for another airport 1040 miles away. The rate of the corporate jet is four times the rate of the single-engine plane. The single-engine plane arrives 8 h after the corporate jet. Find the rate of each plane.

3. ____________

4. ____________

5. The speed of a boat in still water is 28 mph. The boat traveled 70 mi down a river in the same amount of time it traveled 42 mi up the river. Find the rate of the river's current.

6. A jet can fly 550 mph in calm air. Traveling with the wind, the plane can fly 2400 mi in the same amount of time as it flies 2000 mi against the wind. Find the rate of the wind.

5. ____________

6. ____________

7. A motorist drove 270 mi on a four-lane highway before driving 54 mi on a two-lane mountain road. The rate of speed on the four-lane highway was two times the rate on the mountain road. The time spent traveling the 324 mi was 7 h. Find the rate of the motorist on the mountain road.

8. A plane can fly at a rate of 220 mph in calm air. Traveling with the wind, the plane flew 1000 mi in the same amount of time as it flew 760 mi against the wind. Find the rate of the wind.

7. ____________

8. ____________

9. A canoeist can paddle at the rate of 8 mph in calm water. Traveling with the current, the canoe traveled 30 mi in the same amount of time as it traveled 18 mi against the current. Find the rate of the current.

10. A corporate jet can fly at 280 mph in calm air. Flying with the wind the plane can fly 960 mi in the same amount of time as it can fly 720 mi against the wind. Find the rate of the wind.

9. ____________

10. ____________

Name Score

Simplify.

1. $\sqrt{25}$ **2.** $\sqrt{121}$ **3.** $\sqrt{169}$

4. $\sqrt{196}$ **5.** $\sqrt{289}$ **6.** $\sqrt{400}$

7. $\sqrt{48}$ **8.** $\sqrt{20}$ **9.** $\sqrt{45}$

10. $\sqrt{72}$ **11.** $-6\sqrt{32}$ **12.** $9\sqrt{40}$

13. $\sqrt{125}$ **14.** $-7\sqrt{50}$ **15.** $\sqrt{192}$

16. $3\sqrt{144}$ **17.** $\sqrt{150}$ **18.** $\sqrt{900}$

19. $\sqrt{200}$ **20.** $3\sqrt{160}$ **21.** $5\sqrt{98}$

22. $\sqrt{88}$ **23.** $\sqrt{135}$ **24.** $\sqrt{128}$

25. $\sqrt{700}$ **26.** $\sqrt{245}$ **27.** $2\sqrt{96}$

1. ______
2. ______
3. ______
4. ______
5. ______
6. ______
7. ______
8. ______
9. ______
10. ______
11. ______
12. ______
13. ______
14. ______
15. ______
16. ______
17. ______
18. ______
19. ______
20. ______
21. ______
22. ______
23. ______
24. ______
25. ______
26. ______
27. ______

Name Score

Simplify.

1. $\sqrt{x^8}$ **2.** $\sqrt{y^9}$ **3.** $\sqrt{x^{13}}$

4. $\sqrt{x^6y^6}$ **5.** $\sqrt{x^{10}y^4}$ **6.** $\sqrt{9x^6}$

7. $\sqrt{16y^{10}}$ **8.** $\sqrt{x^3y^9}$ **9.** $\sqrt{a^{13}b^3}$

10. $\sqrt{32y^5}$ **11.** $\sqrt{27x^5}$ **12.** $\sqrt{48x^9y^5}$

13. $\sqrt{96x^2y^4}$ **14.** $\sqrt{36a^2b^6}$ **15.** $\sqrt{100x^4y^{16}}$

16. $\sqrt{90x^9y^5}$ **17.** $4\sqrt{16a^6b^3}$ **18.** $3\sqrt{81a^2b^5}$

19. $x\sqrt{x^2y^4}$ **20.** $y\sqrt{x^5y^8}$ **21.** $3\sqrt{9a^3b^2}$

22. $4\sqrt{36a^6b^9}$ **23.** $4y^2\sqrt{32x^2y^3}$ **24.** $\sqrt{64(x+y)^4}$

25. $\sqrt{49(x-3)^2}$ **26.** $\sqrt{4(x-1)^2}$ **27.** $\sqrt{b^2-10b+25}$

1. ______________
2. ______________
3. ______________
4. ______________
5. ______________
6. ______________
7. ______________
8. ______________
9. ______________
10. ______________
11. ______________
12. ______________
13. ______________
14. ______________
15. ______________
16. ______________
17. ______________
18. ______________
19. ______________
20. ______________
21. ______________
22. ______________
23. ______________
24. ______________
25. ______________
26. ______________
27. ______________

Name Score

Simplify.

1. $3\sqrt{3}+2\sqrt{3}$

2. $4\sqrt{2}+7\sqrt{2}$

3. $-2\sqrt{6}+3\sqrt{6}$

4. $-2\sqrt{2}-7\sqrt{2}$

5. $-4\sqrt{5}-6\sqrt{5}$

6. $3\sqrt{y}+7\sqrt{y}$

7. $5\sqrt{x}-9\sqrt{x}$

8. $-4\sqrt{3a}+3\sqrt{3a}$

9. $-4\sqrt{7b}-8\sqrt{7b}$

10. $-6\sqrt{2a}-4\sqrt{2a}$

11. $5x\sqrt{3}-2x\sqrt{3}$

12. $3a\sqrt{5a}-6a\sqrt{5a}$

13. $-4x\sqrt{2y}-3x\sqrt{2y}$

14. $2\sqrt{xy}-10\sqrt{xy}$

15. $-5\sqrt{xy}+7\sqrt{xy}$

16. $4\sqrt{72}-4\sqrt{50}$

17. $6\sqrt{20}-3\sqrt{45}$

18. $4\sqrt{16x}-3\sqrt{25x}$

19. $-2\sqrt{36y}+6\sqrt{64y}$

20. $3\sqrt{16x}-2\sqrt{25x}$

21. $-4\sqrt{36y}+6\sqrt{64y}$

22. $2\sqrt{2y^2}-3\sqrt{32y^2}$

23. $3\sqrt{2}-6\sqrt{2}+7\sqrt{2}$

24. $2\sqrt{5}+7\sqrt{5}-10\sqrt{5}$

25. $2\sqrt{20x}-3\sqrt{45x}+\sqrt{80x}$

26. $3\sqrt{12}-5\sqrt{3}+6\sqrt{27}$

27. $4\sqrt{50}+2\sqrt{32}-\sqrt{18}$

1. ______________
2. ______________
3. ______________
4. ______________
5. ______________
6. ______________
7. ______________
8. ______________
9. ______________
10. ______________
11. ______________
12. ______________
13. ______________
14. ______________
15. ______________
16. ______________
17. ______________
18. ______________
19. ______________
20. ______________
21. ______________
22. ______________
23. ______________
24. ______________
25. ______________
26. ______________
27. ______________

Name Score

Simplify.

1. $\sqrt{3}\sqrt{27}$
2. $\sqrt{5}\sqrt{20}$
3. $\sqrt{z}\sqrt{z}$
4. $\sqrt{xy^5}\sqrt{x^3y}$
5. $\sqrt{a^2b}\sqrt{a^4b^3}$
6. $\sqrt{8a^3b}\sqrt{2ab^5}$
7. $\sqrt{2xy^3}\sqrt{18x^5y}$
8. $\sqrt{5a^6b^3}\sqrt{20a^2b}$
9. $\sqrt{5ab^3}\sqrt{15a^9b^5}$
10. $\sqrt{6}(\sqrt{6}-\sqrt{3})$
11. $4(\sqrt{8}-\sqrt{2})$
12. $\sqrt{2}(\sqrt{2}-\sqrt{8})$
13. $\sqrt{x}(\sqrt{x}+\sqrt{y})$
14. $\sqrt{3}(\sqrt{12}-\sqrt{x})$
15. $\sqrt{8}(\sqrt{y}-\sqrt{2})$
16. $\sqrt{6}(\sqrt{2}-\sqrt{6})$
17. $\sqrt{5}(\sqrt{20}-\sqrt{a})$
18. $(\sqrt{x}-4)^2$
19. $(3\sqrt{x}-y)^2$
20. $\sqrt{2a}(\sqrt{2a}-\sqrt{8b})$
21. $\sqrt{3x}(\sqrt{6x}-\sqrt{3x})$
22. $(2\sqrt{x}-3y)(3\sqrt{x}-5y)$
23. $(4\sqrt{x}+\sqrt{y})(5\sqrt{x}-\sqrt{y})$
24. $(2\sqrt{x}-\sqrt{y})(2\sqrt{x}+\sqrt{y})$
25. $(\sqrt{5x}+y)(\sqrt{5x}-y)$
26. $(3\sqrt{x}+\sqrt{y})(5\sqrt{x}-3\sqrt{y})$
27. $(4\sqrt{x}-3\sqrt{y})(2\sqrt{x}-5\sqrt{y})$

1. ________
2. ________
3. ________
4. ________
5. ________
6. ________
7. ________
8. ________
9. ________
10. ________
11. ________
12. ________
13. ________
14. ________
15. ________
16. ________
17. ________
18. ________
19. ________
20. ________
21. ________
22. ________
23. ________
24. ________
25. ________
26. ________
27. ________

Name Score

Simplify.

1. $\dfrac{\sqrt{50}}{\sqrt{2}}$

2. $\dfrac{\sqrt{54}}{\sqrt{6}}$

3. $\dfrac{\sqrt{80}}{\sqrt{5}}$

4. $\dfrac{\sqrt{162}}{\sqrt{2}}$

5. $\dfrac{\sqrt{108x}}{\sqrt{3x}}$

6. $\dfrac{\sqrt{75x^3}}{\sqrt{3x}}$

7. $\dfrac{\sqrt{24xy^3}}{\sqrt{6xy}}$

8. $\dfrac{1}{\sqrt{5}}$

9. $\dfrac{1}{\sqrt{2}}$

10. $\dfrac{5}{\sqrt{x}}$

11. $\dfrac{6}{\sqrt{3x}}$

12. $\dfrac{\sqrt{48x^4y}}{\sqrt{3x^2y^3}}$

13. $\dfrac{\sqrt{3xy^5}}{\sqrt{27x^3}}$

14. $\dfrac{\sqrt{32xy^5}}{\sqrt{2x^3y}}$

15. $\dfrac{\sqrt{30xy}}{\sqrt{10x^3y^3}}$

16. $\dfrac{1}{\sqrt{3}-1}$

17. $\dfrac{3}{\sqrt{5}-2}$

18. $\dfrac{9}{3+\sqrt{6}}$

19. $\dfrac{6}{\sqrt{2}-5}$

20. $\dfrac{2}{\sqrt{3}-3}$

21. $\dfrac{\sqrt{x}}{\sqrt{x}+\sqrt{y}}$

22. $\dfrac{3\sqrt{3}-6\sqrt{3}}{\sqrt{3}}$

23. $\dfrac{9\sqrt{2}-3\sqrt{2}}{3\sqrt{2}}$

24. $\dfrac{3\sqrt{27}+\sqrt{3}}{\sqrt{12}}$

25. $\dfrac{8x\sqrt{2}-3\sqrt{128}}{\sqrt{32}}$

26. $\dfrac{2y\sqrt{y^3}+\sqrt{y^3}}{\sqrt{y}}$

27. $\dfrac{9a^2\sqrt{a}-3a\sqrt{a^5}}{\sqrt{4a}}$

1. ________
2. ________
3. ________
4. ________
5. ________
6. ________
7. ________
8. ________
9. ________
10. ________
11. ________
12. ________
13. ________
14. ________
15. ________
16. ________
17. ________
18. ________
19. ________
20. ________
21. ________
22. ________
23. ________
24. ________
25. ________
26. ________
27. ________

Name Score

Solve and check.

1. $\sqrt{y}=8$

2. $\sqrt{a}=16$

3. $\sqrt{a}=10$

4. $\sqrt{3x}=3$

5. $\sqrt{2x}=4$

6. $\sqrt{3x}=6$

7. $5-\sqrt{5x}=0$

8. $\sqrt{6x}-12=0$

9. $3-\sqrt{2x}=0$

10. $\sqrt{2x-3}=5$

11. $\sqrt{4x+5}=1$

12. $\sqrt{2x+7}=3$

13. $\sqrt{2x-1}=3$

14. $\sqrt{2x-3}=1$

15. $\sqrt{6x+1}=5$

16. $\sqrt{5x-4}=6$

17. $0=3-\sqrt{2+x}$

18. $0=5-\sqrt{5+x}$

19. $\sqrt{2x}-1=3$

20. $5=7-\sqrt{2c}$

21. $10=3+\sqrt{2x+1}$

22. $\sqrt{2x+5}=0$

23. $\sqrt{3x-2}=0$

24. $0=\sqrt{3x-6}-9$

25. $\sqrt{2x}-4=-6$

26. $\sqrt{2x+5}-3=0$

27. $\sqrt{3x-6}=\sqrt{9x-12}$

1. ____________
2. ____________
3. ____________
4. ____________
5. ____________
6. ____________
7. ____________
8. ____________
9. ____________
10. ____________
11. ____________
12. ____________
13. ____________
14. ____________
15. ____________
16. ____________
17. ____________
18. ____________
19. ____________
20. ____________
21. ____________
22. ____________
23. ____________
24. ____________
25. ____________
26. ____________
27. ____________

Name Score

Solve.

1. Six added to the square root of the product of two and a number is equal to ten. Find the number.

2. The product of a number and the square root of three is equal to the square root of twelve. Find the number.

3. How far would a submarine periscope have to be above the water to locate a ship 4 mi away? The equation for the distance in miles that the lookout can see is $d = 1.4\sqrt{h}$, where h is the height in feet above the surface of the water. Round to the nearest hundredth.

4. How far would a submarine periscope have to be above the water to locate a ship 4.2 mi away? The equation for the distance in miles that the lookout can see is $d = 1.4\sqrt{h}$, where h is the height in feet above the surface of the water. Round to the nearest hundredth.

5. An object is dropped from a helicopter. Find the distance the object has fallen when the speed reaches 96 ft/s. The equation for the distance is $v = \sqrt{64d}$, where v is the speed of the object and d is the distance.

6. An object is dropped from a helicopter. Find the distance the object has fallen when the speed reaches 112 ft/s. The equation for the distance is $v = \sqrt{64d}$, where v is the speed of the object and d is the distance.

7. A stone is dropped from a bridge and hits the water 2 s later. How high is the bridge? The equation for the distance an object falls in T seconds is given by $T = \sqrt{\frac{d}{16}}$, where d is the distance in feet.

8. A stone is dropped into a mine shaft and hits the bottom 4 s later. How deep is the mine shaft? The equation for the distance an object falls in T seconds is given by $T = \sqrt{\frac{d}{16}}$, where d is the distance in feet.

9. Find the length of a pendulum that makes one swing in 4 s. The equation for the time of one swing of the pendulum is given by $T = 2\pi\sqrt{\frac{L}{32}}$, where T is the time in seconds and L is the length in feet. Use 3.14 for π. Round to the nearest hundredth.

10. Find the length of a pendulum that makes one swing in 2.4 s. The equation for the time of one swing of the pendulum is given by $T = 2\pi\sqrt{\frac{L}{32}}$, where T is the time in seconds and L is the length in feet. Use 3.14 for π. Round to the nearest hundredth.

1. __________

2. __________

3. __________

4. __________

5. __________

6. __________

7. __________

8. __________

9. __________

10. __________

Name Score

Solve by factoring.

1. $x^2 - 2x - 24 = 0$
2. $t^2 - 3t - 18 = 0$
3. $z^2 - 6z + 5 = 0$
4. $v^2 - 7v + 6 = 0$
5. $x^2 + 9x + 8 = 0$
6. $v^2 + 10v - 9 = 0$
7. $x^2 - 10x + 25 = 0$
8. $y^2 - 12y + 36 = 0$
9. $10y^2 + 5y = 0$
10. $8x^2 - 12x = 0$
11. $x^2 - 21 = 4x$
12. $t^2 - 20 = t$
13. $2x^2 + x = 10$
14. $4x^2 + 12x = -9$
15. $5r^2 = 12 + 11r$
16. $3x^2 = x + 2$
17. $5y^2 = 11y - 2$
18. $16v^2 - 8v + 1 = 0$
19. $25y^2 - 10y + 1 = 0$
20. $x^2 - 1 = 0$
21. $t^2 - 25 = 0$
22. $16y^2 - 1 = 0$
23. $4z^2 - 25 = 0$
24. $x + 28 = x(x - 2)$
25. $y + 12 = y(y - 3)$
26. $x^2 - 4x - 4 = 3x^2 - 5x - 3$
27. $x^2 + x - 1 = (2x + 1)(x + 2)$

1. ____________
2. ____________
3. ____________
4. ____________
5. ____________
6. ____________
7. ____________
8. ____________
9. ____________
10. ____________
11. ____________
12. ____________
13. ____________
14. ____________
15. ____________
16. ____________
17. ____________
18. ____________
19. ____________
20. ____________
21. ____________
22. ____________
23. ____________
24. ____________
25. ____________
26. ____________
27. ____________

Name Score

Solve by taking square roots.

1. $y^2 = 16$
2. $v^2 - 81 = 0$
3. $z^2 - 144 = 0$
4. $9x^2 - 25 = 0$
5. $16w^2 - 81 = 0$
6. $16w^2 = 25$
7. $4x^2 = 81$
8. $25v^2 - 16 = 0$
9. $36x^2 - 49 = 0$
10. $4x^2 - 9 = 0$
11. $9x^2 - 100 = 0$
12. $x^2 + 9 = 0$
13. $y^2 + 100 = 0$
14. $v^2 - 18 = 0$
15. $x^2 - 50 = 0$
16. $(x+1)^2 = 9$
17. $(y-2)^2 = 36$
18. $3(x+5)^2 = 27$
19. $5(z-2)^2 = 80$
20. $4(x-1)^2 - 25 = 0$
21. $9(y+2)^2 - 100 = 0$
22. $16(w+3)^2 - 49 = 0$
23. $25(y-1)^2 - 36 = 0$
24. $(x-3)^2 - 32 = 0$
25. $(y+4)^2 - 75 = 0$
26. $(x-1)^2 - 50 = 0$
27. $(x+1)^2 - 80 = 0$

1. ______
2. ______
3. ______
4. ______
5. ______
6. ______
7. ______
8. ______
9. ______
10. ______
11. ______
12. ______
13. ______
14. ______
15. ______
16. ______
17. ______
18. ______
19. ______
20. ______
21. ______
22. ______
23. ______
24. ______
25. ______
26. ______
27. ______

Name Score

Solve by completing the square.

1. $x^2 + 6x - 7 = 0$

2. $x^2 + 4x - 21 = 0$

3. $x^2 - 6x - 27 = 0$

4. $w^2 + 2w - 35 = 0$

5. $x^2 = 6x - 9$

6. $v^2 + 2x - 3 = 0$

7. $y^2 - 4y - 7 = 0$

8. $x^2 + 8x = 7$

9. $w^2 - 6w = 1$

10. $z^2 = 4z + 3$

11. $y^2 = 8y - 16$

12. $p^2 + 2p = 3$

13. $r^2 + 3r = 3$

14. $t^2 - 4t = -3$

15. $2y^2 + 5y - 12 = 0$

16. $2x^2 - 5x = -3$

17. $2r^2 + 9t = 5$

18. $3x^2 = 7x - 2$

19. $3y^2 - y = 2$

20. $4x^2 + 7x - 2 = 0$

21. $3y^2 - 4y = 4$

22. $3t^2 - 2t = 5$

23. $4v^2 - 3v = 1$

24. $3t^2 = 2t + 1$

25. $6t^2 - 5t + 1 = 0$

26. $4z^2 - 8z - 5 = 0$

27. $6z^2 = z + 2$

1. ____________
2. ____________
3. ____________
4. ____________
5. ____________
6. ____________
7. ____________
8. ____________
9. ____________
10. ____________
11. ____________
12. ____________
13. ____________
14. ____________
15. ____________
16. ____________
17. ____________
18. ____________
19. ____________
20. ____________
21. ____________
22. ____________
23. ____________
24. ____________
25. ____________
26. ____________
27. ____________

Name ____________ Score ____________

Solve by using the quadratic formula.

1. $x^2 - 3x - 4 = 0$
2. $y^2 + 6y + 5 = 0$
3. $z^2 - 3z - 10 = 0$
4. $v^2 - 6v + 5 = 0$
5. $z^2 + 5z - 6 = 0$
6. $t^2 - t - 20 = 0$
7. $x^2 - x - 6 = 0$
8. $y^2 = 8y + 9$
9. $w^2 = 6w + 16$
10. $r^2 = 9 - 8r$
11. $z^2 = 10 - 9z$
12. $2t^2 - t - 10 = 0$
13. $x^2 - 6x + 6 = 0$
14. $x^2 + 6x + 2 = 0$
15. $p^2 - 4p = 0$
16. $3v^2 - v = 0$
17. $9t^2 - 25 = 0$
18. $16s^2 - 49 = 0$
19. $3y^2 + y = 14$
20. $2x^2 - 7x = 4$
21. $3x^2 = -2x + 8$
22. $2x^2 + x = 10$
23. $5x^2 - 13x = 6$
24. $3v^2 = -v + 2$
25. $t^2 - t - 3 = 0$
26. $x^2 + 6x = 4$
27. $t^2 + 5t - 1 = 0$

1. ____________
2. ____________
3. ____________
4. ____________
5. ____________
6. ____________
7. ____________
8. ____________
9. ____________
10. ____________
11. ____________
12. ____________
13. ____________
14. ____________
15. ____________
16. ____________
17. ____________
18. ____________
19. ____________
20. ____________
21. ____________
22. ____________
23. ____________
24. ____________
25. ____________
26. ____________
27. ____________

Name Score

Write each expression in terms of *i*.

1. $-3\sqrt{-27}$
2. $\sqrt{-8}$
3. $\sqrt{-12}$
4. $\sqrt{-200}$
5. $\sqrt{-128}$
6. $\sqrt{-144}$
7. $\sqrt{-108}$
8. $\sqrt{-180}$

1. ______
2. ______
3. ______
4. ______
5. ______
6. ______
7. ______
8. ______

Write each number in standard form.

9. $\sqrt{9}+\sqrt{-9}$
10. $\sqrt{16}+\sqrt{-25}$
11. $\sqrt{36}+\sqrt{-16}$
12. $\sqrt{49}+\sqrt{-100}$
13. $\sqrt{48}-\sqrt{-27}$
14. $\sqrt{24}-\sqrt{-36}$
15. $\sqrt{50}-\sqrt{-98}$
16. $\sqrt{150}-\sqrt{-27}$

9. ______
10. ______
11. ______
12. ______
13. ______
14. ______
15. ______
16. ______

Name Score

Add or subtract.

1. $-4i+2i$

2. $10i+11i$

3. $3i-(-5i)$

4. $(5-8i)+(6+2i)$

5. $(4-6i)+(7-3i)$

6. $(4+6i)+(8-3i)$

7. $(7-4i)+(5+3i)$

8. $(4-6i)+(9-i)$

9. $(9-2i)+i$

10. $(8-5i)+2i$

11. $(6-4i)+3i$

12. $(7-6i)+5i$

13. $(5+4i)+(-5-4i)$

14. $(7-5i)+(-10+2i)$

15. $(8+3i)+(-6-4i)$

16. $(9-2i)+(-4+4i)$

17. $(3+8i)-(10-9i)$

18. $(-5+7i)-(3+2i)$

1. ____________
2. ____________
3. ____________
4. ____________
5. ____________
6. ____________
7. ____________
8. ____________
9. ____________
10. ____________
11. ____________
12. ____________
13. ____________
14. ____________
15. ____________
16. ____________
17. ____________
18. ____________

Name Score

Multiply.

1. $(5i)(-8i)$

2. $(-5i)(-3i)$

3. $(-8i)(20)$

4. $(-4i)(-8i)$

5. $3i(5+i)$

6. $4i(2+3i)$

7. $(5-2i)(3+i)$

8. $(4+3i)(3+i)$

9. $(2-i)(5-i)$

10. $(3-4i)(2-i)$

11. $(2+i)(5-3i)$

12. $(3-i)(3-6i)$

13. $(5+4i)(2+i)$

14. $(5-i)^2$

15. $(5+2i)^2$

16. $(6+2i)(6-2i)$

17. $(3-5i)(3+5i)$

18. $(2-i)(2+i)$

1. ______________
2. ______________
3. ______________
4. ______________
5. ______________
6. ______________
7. ______________
8. ______________
9. ______________
10. ______________
11. ______________
12. ______________
13. ______________
14. ______________
15. ______________
16. ______________
17. ______________
18. ______________

Name Score

Divide.

1. $\frac{4}{i}$

2. $\frac{2}{3i}$

3. $\frac{3}{4i}$

4. $\frac{4-6i}{-2i}$

5. $\frac{15+2i}{-5i}$

6. $\frac{5}{4+i}$

7. $\frac{8}{3+i}$

8. $\frac{4}{4-i}$

9. $\frac{6}{5-i}$

10. $\frac{1-5i}{5+i}$

11. $\frac{1-4i}{4+i}$

12. $\frac{4+20i}{5+i}$

13. $\frac{6-3i}{3-i}$

14. $\frac{4+8i}{2-i}$

15. $\frac{3-2i}{2+i}$

16. $\frac{4+3i}{2-i}$

17. $\frac{2+5i}{1-i}$

18. $\frac{6+5i}{2-i}$

1. ______________

2. ______________

3. ______________

4. ______________

5. ______________

6. ______________

7. ______________

8. ______________

9. ______________

10. ______________

11. ______________

12. ______________

13. ______________

14. ______________

15. ______________

16. ______________

17. ______________

18. ______________

Name Score

Solve.

1. $x^2 + 9 = 0$

2. $x^2 + 64 = 0$

3. $6x^2 + 54 = 0$

4. $11x^2 + 539 = 0$

5. $x^2 - 6x + 13 = 0$

6. $x^2 + 4x + 13 = 0$

7. $(x+1)^2 + 9 = 7$

8. $(x+3)^2 + 20 = 4$

9. $x^2 - 2x + 6 = 0$

10. $x^2 + 8x = -17$

11. $(x+1)^2 + 36 = 0$

12. $x^2 + 2x + 4 = 0$

13. $(x-5)^2 + 25 = 0$

14. $(x+3)^2 + 64 = 0$

15. $(2x-1)^2 + 55 = 6$

16. $(3x+2)^2 + 29 = 28$

1. ______________
2. ______________
3. ______________
4. ______________
5. ______________
6. ______________
7. ______________
8. ______________
9. ______________
10. ______________
11. ______________
12. ______________
13. ______________
14. ______________
15. ______________
16. ______________

Name Score

Graph.

1. $y = x^2 + 1$

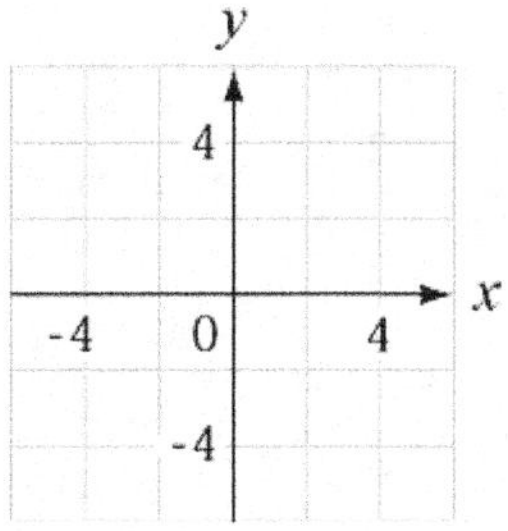

2. $y = -x^2 + 3$

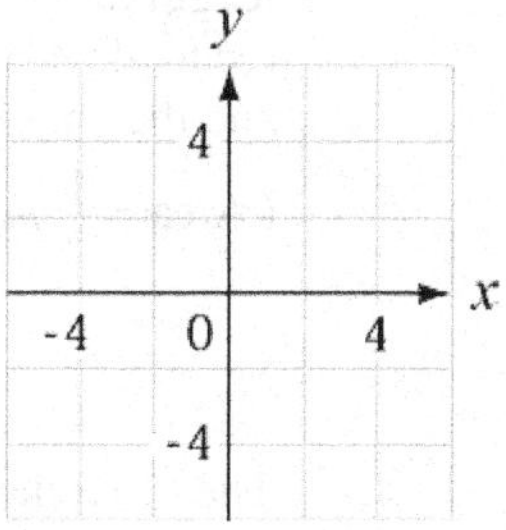

3. $y = -2x^2$

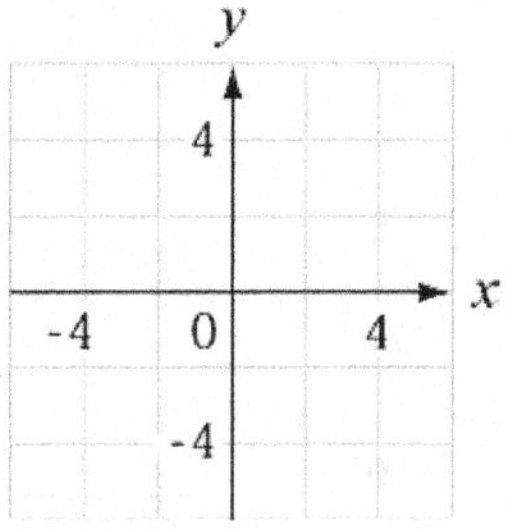

4. $y = x^2 - 2x$

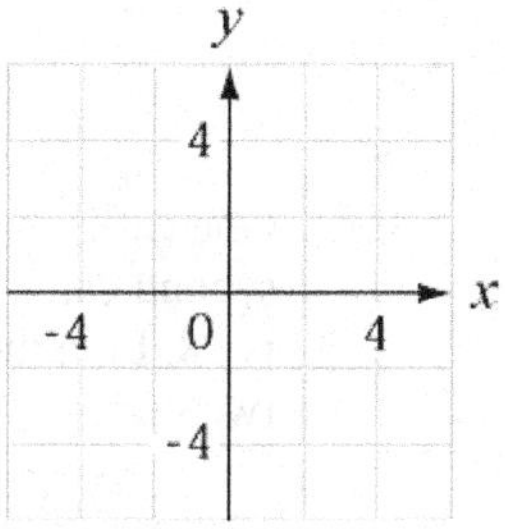

5. $y = \frac{1}{2}x^2 - 1$

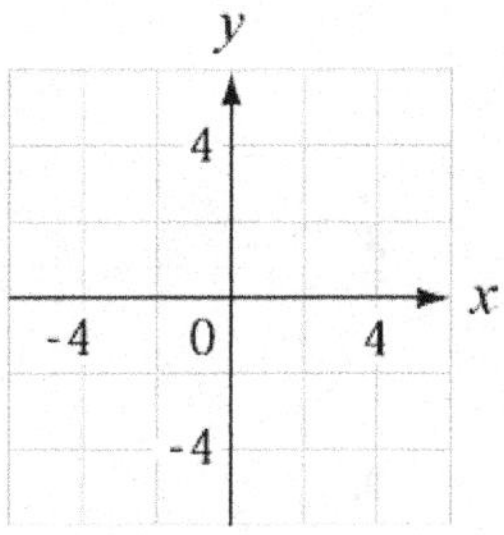

6. $y = x^2 - 4x + 4$

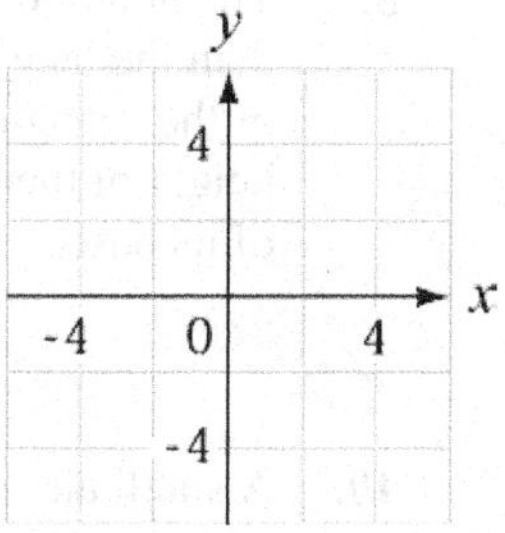

7. $y = x^2 - 3x + 2$

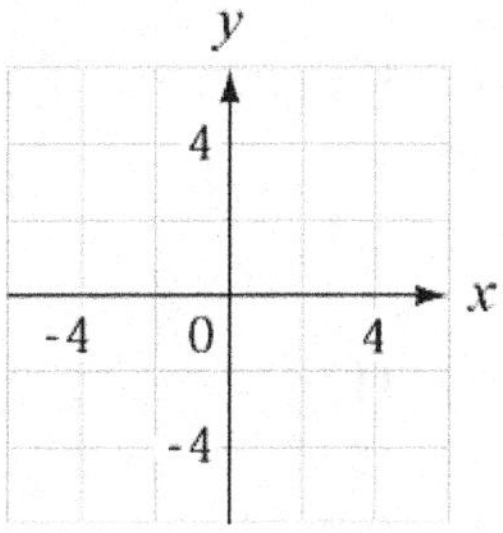

8. $y = x^2 - x - 2$

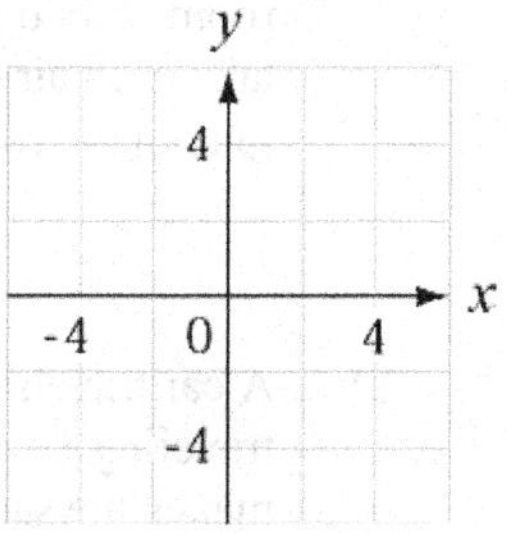

Name Score

Solve.

1. The length of a rectangle is three times the width. The area of the rectangle is 48 ft^2. Find the length and width. (Area $= l \cdot w$)

2. The height of a triangle is twice the length of the base. The area of the rectangle is 36 m^2. Find the height and the length of the base of the triangle. (Area $= \frac{1}{2}bh$)

3. The sum of the squares of two consecutive positive even integers is fifty-two. Find the two integers.

4. The difference between the squares of two consecutive odd integers is thirty-two. Find the two integers.

5. The sum of the squares of three consecutive odd integers is eleven. Find the three integers.

6. One car is three years older than a second car. The product of their ages is 18. Find the present ages of the two cars.

7. The length of a rectangle is 2 ft less than three times the width. The area of the rectangle is 40 ft^2. Find the length and width of the rectangle.

8. The height of a triangle is 3 m less than the length of the base. The area of the triangle is 20 m^2. Find the height of the triangle and the length of its base.

9. An old pump requires 9 h longer to empty a pool than does a new pump. With both pumps working, the pool can be emptied in 6 h. Find the time required for the new pump, working alone, to empty the pool.

10. A small air conditioner requires 16 min longer to cool a room 4° than does a larger air conditioner. Working together, the two air conditioners can cool the room 4° in 6 min. How long would it take each air conditioner, working alone, to cool the room 4°?

11. An old copier takes 15 min longer to copy a report than does a second, new model. With both copiers working, the report can be printed in 18 min. How long would it take each copier, working alone, to copy the report?

12. A car travels 240 mi. A second car, traveling 12 mph faster than the first, makes the same trip in 1 h less time. Find the speed of each car.

1. ______________
2. ______________
3. ______________
4. ______________
5. ______________
6. ______________
7. ______________
8. ______________
9. ______________
10. ______________
11. ______________
12. ______________

CHAPTER 1

Objective 1.1.1

1. >	**2.** <	**3.** >	**4.** <	**5.** >	**6.** <	**7.** False
8. False	**9.** True	**10.** False	**11.** False	**12.** False	**13.** False	**14.** False
15. True	**16.** {1, 2, 3}	**17.** {–3, –2, –1}	**18.** {5, 4, 3, 2, 1}	**19.** –6, –3	**20.** –5, –1, 3	
21. –20, 0	**22.** 2, 3, 7, 18	**23.** –26, –19, –7, –5, 2, 6		**24.** 0, 3, 6		

Objective 1.1.2

1. –7	**2.** 4	**3.** 5	**4.** –34	**5.** 28	**6.** –66	**7.** –3
8. –7	**9.** 5	**10.** 13	**11.** –4	**12.** 15	**13.** 16	**14.** –19
15. –19	**16.** <	**17.** >	**18.** >	**19.** <	**20.** <	**21.** <
22. >	**23.** <	**24.** >				

Objective 1.2.1

1. –2	**2.** –2	**3.** –10	**4.** 7	**5.** 2	**6.** –18	**7.** –7
8. –14	**9.** –5	**10.** –6	**11.** 4	**12.** –4	**13.** –5	**14.** –25
15. –5	**16.** 10	**17.** 36	**18.** –4	**19.** –92	**20.** –39	**21.** –9
22. –1	**23.** –17	**24.** –27	**25.** –3	**26.** –23	**27.** –9	

Objective 1.2.2

1. 6	**2.** 10	**3.** –10	**4.** –9	**5.** –12	**6.** 10	**7.** 4
8. –8	**9.** 4	**10.** –28	**11.** –8	**12.** –20	**13.** –7	**14.** 12
15. –7	**16.** –8	**17.** 8	**18.** 1	**19.** –9	**20.** 54	**21.** –129
22. 14	**23.** –5	**24.** –24	**25.** 64	**26.** –83	**27.** 6	

Objective 1.2.3

1. 44	**2.** –8	**3.** –12	**4.** 14	**5.** –16	**6.** –24	**7.** –42
8. –20	**9.** 28	**10.** 18	**11.** 0	**12.** –63	**13.** –90	**14.** –57
15. 52	**16.** –108	**17.** –162	**18.** 192	**19.** –180	**20.** –42	**21.** –120
22. –108	**23.** 48	**24.** –210	**25.** 56	**26.** –180	**27.** –48	

Objective 1.2.4

1. –3	**2.** –6	**3.** 8	**4.** –1	**5.** –7	**6.** –5	**7.** –6
8. –8	**9.** 9	**10.** 6	**11.** 7	**12.** –10	**13.** 5	**14.** 12
15. 12	**16.** 15	**17.** 15	**18.** –17	**19.** –8	**20.** –16	**21.** –16
22. –7	**23.** 14	**24.** –41	**25.** 15	**26.** 9	**27.** –14	

Objective 1.2.5

1. 9250 m	**2.** 107.2°F	**3.** 107°C	**4.** 379°C	**5.** 7°F	**6.** 5°C	**7.** –3°C
8. –1°C	**9.** 14 shots	**10.** 20 shots				

Objective 1.3.1

1. 0.375	**2.** $0.\overline{7}$	**3.** $0.08\overline{3}$	**4.** $0.4\overline{6}$	**5.** 0.3125	**6.** 0.15	**7.** 0.45
8. 0.95	**9.** 0.12	**10.** 0.28	**11.** $0.3\overline{6}$	**12.** 0.5		

Answers to AIM for Success Practice Sheets

Objective 1.3.2

1. $-\frac{1}{20}$	**2.** $-\frac{1}{20}$	**3.** $\frac{1}{24}$	**4.** $-\frac{16}{45}$	**5.** $-\frac{9}{16}$	**6.** $\frac{1}{24}$	**7.** $\frac{7}{9}$
8. –3.337	**9.** 9.086	**10.** $\frac{2}{3}$	**11.** $-\frac{35}{24}$	**12.** –57.49	**13.** $\frac{7}{12}$	**14.** $\frac{4}{9}$
15. $-\frac{29}{48}$	**16.** $-1\frac{1}{9}$	**17.** $\frac{11}{24}$	**18.** $\frac{29}{30}$	**19.** $-\frac{11}{36}$	**20.** $\frac{11}{24}$	**21.** –4.88
22. 0	**23.** $-\frac{1}{8}$	**24.** $\frac{5}{4}$				

Objective 1.3.3

1. $-\frac{1}{5}$	**2.** $\frac{3}{8}$	**3.** $\frac{1}{20}$	**4.** $-\frac{9}{50}$	**5.** $\frac{36}{77}$	**6.** $-\frac{4}{15}$	**7.** 4.26
8. –3.24	**9.** 2.136	**10.** $1\frac{2}{3}$	**11.** $-1\frac{1}{2}$	**12.** $-\frac{5}{6}$	**13.** $-2\frac{1}{12}$	**14.** $1\frac{1}{8}$
15. $-\frac{3}{4}$	**16.** $-\frac{3}{5}$	**17.** $1\frac{1}{3}$	**18.** $1\frac{1}{3}$	**19.** $-\frac{1}{12}$	**20.** –4.5	**21.** $\frac{1}{2}$
22. –0.163	**23.** 0.014	**24.** $-\frac{23}{12}$				

Objective 1.3.4

1. $\frac{3}{10}$; 0.30	**2.** $\frac{11}{25}$; 0.44	**3.** $1\frac{1}{10}$; 1.10	**4.** $1\frac{2}{5}$; 1.40	**5.** $\frac{21}{100}$; 0.21	**6.** $\frac{3}{20}$; 0.15	**7.** $\frac{2}{9}$
8. $\frac{11}{250}$	**9.** $\frac{137}{400}$	**10.** 0.038	**11.** 0.1617	**12.** 0.3207	**13.** 118%	**14.** 205.4%
15. 0.9%	**16.** 60%	**17.** 133.3%	**18.** 375%	**19.** 52%	**20.** $21\frac{3}{7}$%	**21.** 312.5%

Objective 1.4.1

1. 16	**2.** 125	**3.** –25	**4.** –81	**5.** 16	**6.** –27	**7.** 25
8. 16	**9.** –8	**10.** –27	**11.** 1728	**12.** 256	**13.** –4	**14.** 8
15. 400	**16.** –320	**17.** 576	**18.** –48	**19.** –12	**20.** 24	**21.** –180
22. 216	**23.** –432	**24.** –72				

Objective 1.4.2

1. 3	**2.** 14	**3.** 8	**4.** 11	**5.** 20	**6.** 19	**7.** 5
8. 5	**9.** –16	**10.** –5	**11.** –13	**12.** 0	**13.** $-\frac{3}{2}$	**14.** 6
15. 10	**16.** –2	**17.** 6	**18.** 1	**19.** –49	**20.** 3	**21.** 8
22. 9	**23.** 8	**24.** 1.305				

Objective 1.5.1

1. 38°	**2.** 11°	**3.** 81°	**4.** 0°	**5.** 128°	**6.** 101°	**7.** 171°
8. 69°	**9.** 1°	**10.** 90°	**11.** 149°	**12.** 113°	**13.** 63°	**14.** 116°
15. 84°	**16.** 268°					

Objective 1.5.2

1. 14 ft 10 in. **2.** 18 ft 6 in. **3.** 24 ft 8 in. **4.** 148 cm **5.** 36 m **6.** 13.06 cm **7.** 144.44 cm
8. 5.024 m **9.** \$114.60 **10.** \$48.80

Objective 1.5.3

1. 65 ft^2 **2.** 330.17 cm^2 **3.** 153 ft^2 **4.** 78.5 $in.^2$ **5.** 181.3664 m^2 **6.** 27.04 m^2 **7.** 8.41 yd^2
8. 56 cm^2 **9.** 171 $in.^2$ **10.** 16.6106 m^2 **11.** 25 yd^2 **12.** \$1187

CHAPTER 2

Objective 2.1.1

1. 13 **2.** 2 **3.** 25 **4.** 8 **5.** 3 **6.** 22 **7.** 49
8. 7 **9.** 2 **10.** 25 **11.** –10 **12.** 1 **13.** –32 **14.** 17
15. 15 **16.** $-\frac{1}{2}$ **17.** –7 **18.** 1 **19.** 22 **20.** –3 **21.** –8
22. 4 **23.** 58 **24.** 1

Objective 2.2.1

1. 6 **2.** 8 **3.** 1 **4.** $-x$ **5.** 0 **6.** wz **7.** 20
8. 1 **9.** 0 **10.** 3 **11.** The Inverse Property of Addition **12.** The Distributive Property
13. The Multiplication Property of Zero **14.** The Multiplication Property of One
15. The Associative Property of Addition **16.** The Addition Property of Zero
17. The Associative Property of Multiplication **18.** The Inverse Property of Multiplication
19. The Commutative Property of Addition **20.** The Commutative Property of Multiplication

Objective 2.2.2

1. $11x$ **2.** $21x$ **3.** $5a$ **4.** $7a$ **5.** $2ab$ **6.** $5xy$ **7.** $-2xy$
8. $-4xy$ **9.** $5ab$ **10.** 0 **11.** $-\frac{7}{12}x$ **12.** $-\frac{2}{5}x$ **13.** $-\frac{1}{3}x^2$ **14.** $\frac{1}{8}x$
15. $-\frac{17}{12}x^2$ **16.** $8a$ **17.** $-2a$ **18.** $-y^2$ **19.** $x+2y$ **20.** $7y-5x$ **21.** $7x$
22. $6a-12b$ **23.** $14a-13b$ **24.** $7y-11x$ **25.** $-2x^2-x$

Objective 2.2.3

1. $10x$ **2.** $24x$ **3.** $-6a$ **4.** $12y$ **5.** $56y$ **6.** $-12x^2$ **7.** $-63x^2$
8. x^2 **9.** x^2 **10.** a **11.** x **12.** x **13.** a **14.** a
15. x **16.** x **17.** y **18.** n **19.** $3x$ **20.** $4x$ **21.** $-4x$
22. $-3x$ **23.** $-28a^2$ **24.** $-15x^2$ **25.** $5y$ **26.** $8y$ **27.** $4y$

Objective 2.2.4

1. $-x-5$ **2.** $6x-3$ **3.** $8x-12$ **4.** $-3a-6$ **5.** $-4a-48$ **6.** $-4y+10$ **7.** $-12y+16$
8. $-6x^2-36$ **9.** $-20x^2-8$ **10.** $-10y^2+15$ **11.** $-21y^2+70$ **12.** $-6x^2-6y^2$ **13.** $-5x^2+5y^2$ **14.** $-3x^2+6y^2$
15. $-10a^2+16b^2$ **16.** $6x^2+6x-42$ **17.** $3x^2-6x-15$ **18.** $-2y^2-6y+10$
19. $-3a^2-3a-12$ **20.** $14x^2-28x-42$ **21.** $6x^2-8x+2$ **22.** $2x^2+4xy-8y^2$
23. $6x^2-9xy-12y^2$ **24.** $-2a^2-4a+3$ **25.** $9b^2+5b-8$

Objective 2.2.5

1. $-x-21$	**2.** $-a-1$	**3.** $13-10x$	**4.** $17-6x$	**5.** $-2-5y$	**6.** $7n-6$	**7.** $4x-10$
8. $-2x+26$	**9.** $x-17$	**10.** $4y+2$	**11.** $12y-33$	**12.** $-x+5y$	**13.** $2a+5b$	**14.** $-6x+18$
15. $20x+90$	**16.** $3x-18$	**17.** $4x-48$	**18.** $-4x+16$	**19.** $10x+5$	**20.** $-5x+30$	**21.** $22x-48$
22. $-4x+49$	**23.** $a+2b$	**24.** $8b$	**25.** $13x-24y$	**26.** $22x-y$		

Objective 2.3.1

1. $6+x$	**2.** $y-10$	**3.** $12-x$	**4.** $t+6$	**5.** $a-1$	**6.** $y-5$	**7.** $\frac{z}{8}$
8. $11m$	**9.** x^2+10	**10.** $15-x^3$	**11.** $5(n+8)$	**12.** $\frac{2}{3}x+6$	**13.** $x+3x$	**14.** $\frac{-6}{y}$
15. $-2t$	**16.** $2(x+8)$	**17.** $5(y+6)$	**18.** $\frac{10}{y+5}$	**19.** $\frac{1}{4}y^2+12$	**20.** $-1m+16$	**21.** $x+\frac{x}{2}$
22. $\frac{3}{5}(10w)$	**23.** $y-2y$	**24.** $7(x+3)$				

Objective 2.3.2

1. $4n-n;\ 3n$ **2.** $5n+n;\ 6n$ **3.** $(n+12)-3;\ n+9$ **4.** $n+(11-n);\ 11$ **5.** $n+(n+7);\ 2n+7$

6. $n-(n+5);\ -5$ **7.** $\frac{2}{5}n+\frac{3}{10}n;\ \frac{7}{10}n$ **8.** $\frac{3}{8}n-\frac{1}{4}n;\ \frac{1}{8}n$ **9.** $(n+9)-3;\ n+6$

10. $(n+3)+4;\ n+7$ **11.** $4(5n+6);\ 20n+24$ **12.** $11n-7n;\ 4n$ **13.** $(n+10)+12;\ n+22$

14. $15-(n+5);\ 10-n$ **15.** $3(n+n+1);\ 6n+3$ **16.** $(n+n+1)+9;\ 2n+10$ **17.** $\frac{1}{2}(n+n+2);\ n+1$

18. $(n-9)+(n+20);\ 2n+11$ **19.** $5n-n;\ 4n$ **20.** $6(n^2+4);\ 6n^2+24$

Objective 2.3.3

1. gallons of paint in first container: g; gallons of paint in the second container: $10-g$
2. length of first piece: L; length of second piece: $35-L$ **3.** speed of corporate jet: s; speed of commercial jet: $3s$
4. amount at 3.5%: A; amount at 4%: $4200-A$ **5.** height of triangle: h; base of triangle: $h+4$
6. width: w; length: $4w$ **7.** length of shorter piece: L; length of longer piece: $8-L$
8. width of rectangle: w; length of rectangle: $2w-3$ **9.** number of apple trees: N; number of cherry trees: $\frac{1}{5}N$
10. number of full-time faculty: F; number of adjunct faculty: $F+206$

CHAPTER 3

Objective 3.1.1

1. Yes	**2.** Yes	**3.** No	**4.** Yes	**5.** Yes	**6.** Yes	**7.** No
8. No	**9.** Yes	**10.** No	**11.** Yes	**12.** Yes	**13.** Yes	**14.** Yes
15. No	**16.** Yes	**17.** Yes	**18.** No	**19.** Yes	**20.** Yes	**21.** No
22. Yes	**23.** Yes	**24.** No				

Objective 3.1.2

1. 5	**2.** 3	**3.** 15	**4.** 14	**5.** 9	**6.** –5	**7.** –8
8. 2	**9.** 3	**10.** 4	**11.** 2	**12.** 0	**13.** –5	**14.** 2
15. –6	**16.** 2	**17.** 6	**18.** –8	**19.** 10	**20.** 15	**21.** –6
22. 16	**23.** –1	**24.** $\frac{1}{4}$	**25.** $\frac{3}{5}$	**26.** $-\frac{5}{6}$	**27.** $-\frac{7}{6}$	

Objective 3.1.3

1. 3	**2.** 6	**3.** –5	**4.** –5	**5.** –3	**6.** 12	**7.** 7
8. 6	**9.** –7	**10.** 3	**11.** –6	**12.** –4	**13.** 12	**14.** 20
15. –12	**16.** –15	**17.** 14	**18.** –20	**19.** –16	**20.** 8	**21.** –6
22. –30	**23.** 12	**24.** 16	**25.** $\frac{3}{2}$	**26.** 2	**27.** 5	

Objective 3.2.1

1. 16%	**2.** 20%	**3.** 4.8	**4.** 27	**5.** 400	**6.** 32	**7.** 6
8. 40%	**9.** 40	**10.** 64	**11.** 500%	**12.** 400%	**13.** 7.2	**14.** 3.78
15. 18.48	**16.** 20%	**17.** 52 cars	**18.** 140 employees	**19.** 600 computers		**20.** 20%
21. 57,600 building permits						

Objective 3.2.2

1. 9 miles per minute	**2.** 6 h	**3.** 4 h	**4.** 75 s	**5.** 7 mi	**6.** 4 h
7. 54 min	**8.** 3.4 mi				

Objective 3.3.1

1. 6	**2.** 7	**3.** 4	**4.** 7	**5.** –3	**6.** –5	**7.** –7
8. 2	**9.** 2	**10.** 1	**11.** 5	**12.** 6	**13.** 2	**14.** 2
15. 7	**16.** –4	**17.** –7	**18.** 4	**19.** 9	**20.** 2	**21.** 2
22. 3	**23.** 1	**24.** 2	**25.** 6	**26.** 4	**27.** –9	

Objective 3.3.2

1. 7	**2.** 4	**3.** 8	**4.** 5	**5.** 4	**6.** –2	**7.** –3
8. –2	**9.** –3	**10.** 2	**11.** –3	**12.** –2	**13.** –2	**14.** –4
15. –6	**16.** 0	**17.** 0	**18.** –1	**19.** –2	**20.** 2	**21.** –1
22. –3	**23.** 6	**24.** 10	**25.** –2	**26.** –3	**27.** 6	

Objective 3.3.3

1. 2	**2.** 3	**3.** –6	**4.** 7	**5.** –1	**6.** 1	**7.** –2
8. 3	**9.** 2	**10.** –2	**11.** –7	**12.** 6	**13.** 8	**14.** –3
15. 0	**16.** –1	**17.** –12	**18.** 2	**19.** –3	**20.** 4	**21.** –3
22. 2	**23.** 2	**24.** 9	**25.** –4	**26.** –2	**27.** 0	

Objective 3.3.4

1. \$96	**2.** 42	**3.** 12,320 cm^3	**4.** 15 cm	**5.** 6 ft from the 36-lb force
6. 6 ft from the 50-lb force	**7.** 48 lb	**8.** 8 lb	**9.** 51.25°C	**10.** 60.8°C

Objective 3.4.1

1. 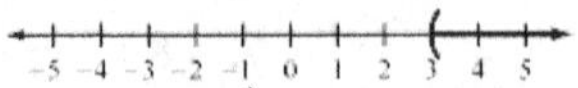
2. 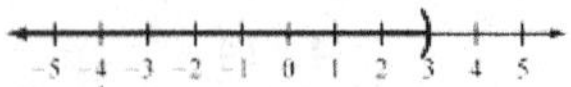
3. 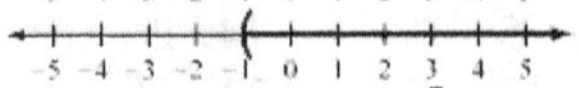
4. 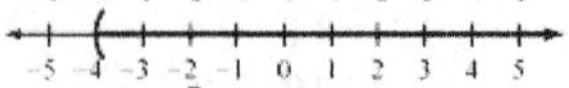
5. 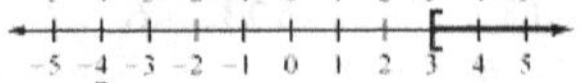
6. 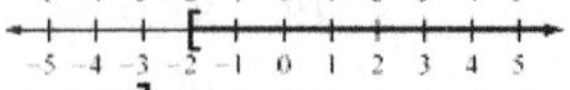
7. 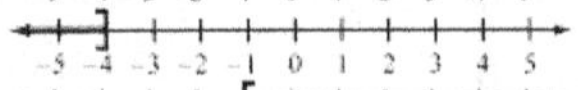
8. 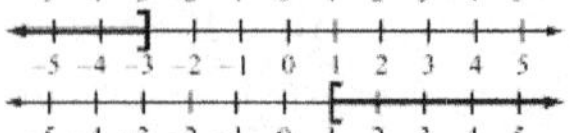
9. 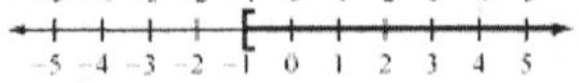
10. 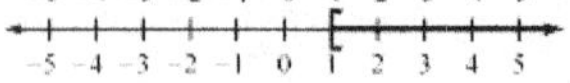

11. $y \geq -10$ **12.** $x \geq -18$ **13.** $x < 9$ **14.** $x \geq 2$ **15.** $x < -8$ **16.** $x < -6$ **17.** $x \geq -2$

18. $d < \frac{1}{2}$ **19.** $y \geq -\frac{3}{8}$ **20.** $b \geq \frac{7}{10}$ **21.** $N \leq 6.4$ **22.** $x < 4$

Objective 3.4.2

1. 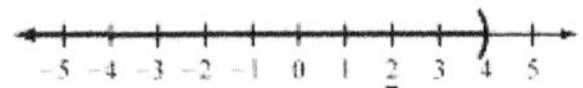
2. 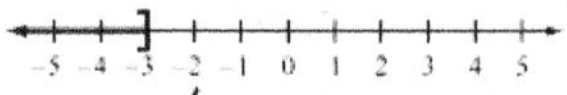
3. 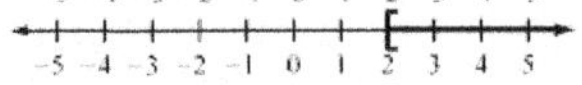
4. 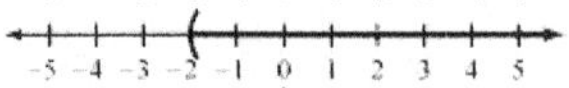
5. 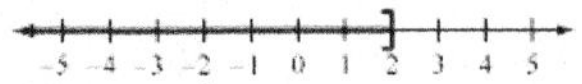
6. 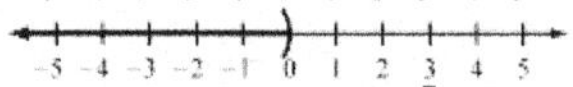
7. 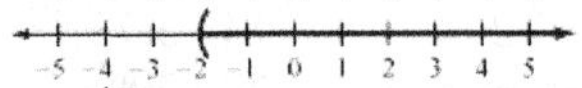
8. 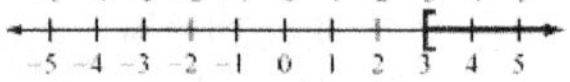
9. 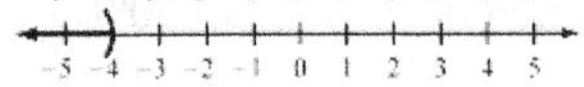
10. 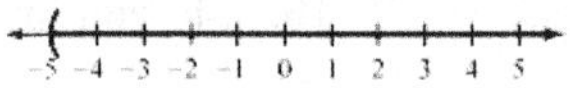

11. $y \leq -4$ **12.** $x < 4$ **13.** $x \leq 16$ **14.** $b \leq 9$ **15.** $x \leq 20$ **16.** $N < \frac{2}{3}$ **17.** $x > \frac{2}{5}$

18. $y \leq \frac{4}{3}$ **19.** $b \leq 22$ **20.** $y \leq -4$ **21.** $x \leq -2$ **22.** $d < -4$

Objective 3.4.3

1. $x < 2$ **2.** $x < 3$ **3.** $y < -4$ **4.** $x \leq 3$ **5.** $N \geq 4$ **6.** $x < -70$ **7.** $b > -5$

8. $x > 5$ **9.** $x > -5$ **10.** $x > 4$ **11.** $x > -7$ **12.** $x > -6$ **13.** $x \leq \frac{1}{2}$ **14.** $x < 2$

15. $x \geq -1$ **16.** $x < 25$ **17.** $d < -7$ **18.** $x > -2$ **19.** $d \geq 2$ **20.** $x > 7$ **21.** $N \leq \frac{1}{2}$

22. $x \leq -7$ **23.** $N \leq 1$ **24.** $x \leq 2$

CHAPTER 4

Objective 4.1.1

1. $n + 9 = 15; n = 6$ **2.** $n + 2 = 11; n = 9$ **3.** $n - 6 = 10; n = 16$ **4.** $\frac{1}{2}n = 9; n = 18$

5. $4n + 5 = 13; n = 2$ **6.** $2(n + 3) = 12; n = 3$ **7.** $3n - 8 = 7; n = 5$ **8.** $\frac{n}{6} = -2; n = -12$

9. $7n + 3 = -18; n = -3$ **10.** $5n + n = 24; n = 4$ **11.** $6n - 3n = 12; n = 4$ **12.** $30 = 5 - 5n; n = -5$
13. $9 = 2n + 1; n = 4$ **14.** $2n = 4(21 - n)$; 7 and 14 **15.** $5n = 2n + 18; n = 6$ **16.** $4(n + 3) = 5n - 2; n = 14$

Objective 4.1.2

1. \$14 **2.** \$80 **3.** \$3200 **4.** \$27,000 **5.** 960 **6.** 80°, 40° **7.** 2 h
8. 40 mph **9.** 14 ft **10.** 1450 rpm

Answers to AIM for Success Practice Sheets

Objective 4.2.1

1. length is 54 m; width is 18 m **2.** 10 ft, 13, ft, 19 ft **3.** length is 27 m; width is 11 m **4.** 40 ft, 80 ft, 60 ft
5. 60 cm, 45 cm, 60 cm **6.** 60 cm, 30 cm, and 90 cm **7.** length is 33 m; width is 26 m
8. length is 40 ft; width is 24 ft **9.** 20 ft, 20 ft, 13 ft **10.** 4 m

Objective 4.2.2

1. 69° **2.** 0° **3.** 81° **4.** 101° **5.** 12° **6.** 14° **7.** 47°
8. 5 **9.** $a = 123°$, $b = 57°$ **10.** $a = 39°$, $b = 141°$ **11.** 20° **12.** 25°

Objective 4.2.3

1. 50° **2.** 37° **3.** 55° **4.** 69° **5.** 14° **6.** 29° **7.** 30°
8. 57°

Objective 4.3.1

1. \$6.75 **2.** \$25 **3.** \$10 **4.** 40% **5.** \$4.26 **6.** \$32.40 **7.** 50%
8. 50% **9.** $33\frac{1}{3}\%$ **10.** 45% **11.** 80% **12.** \$144

Objective 4.3.2

1. \$160 **2.** \$24 **3.** \$40 **4.** \$70 **5.** 20% **6.** 25% **7.** \$40
8. \$150 **9.** 15% **10.** \$17.40 **11.** \$36 **12.** $22\frac{1}{2}\%$

Objective 4.4.1

1. \$3000 at 10%; \$5000 at 12% **2.** \$5000 at 9%; \$9000 at 15% **3.** \$6000 **4.** \$4800 at 18%; \$7200 at 12%
5. \$4000 at 15%; \$3000 at 20% **6.** \$6000 **7.** \$2,000,000 **8.** \$30,000

Objective 4.5.1

1. 12 lb at \$4.50/lb and 8 lb at \$3/lb **2.** 12 lb **3.** 18 kg at \$3/kg and 12 kg at \$4/kg **4.** 60 oz
5. 30 lb **6.** \$3.20 **7.** 15 kg of peanuts and 30 kg of cashews **8.** \$0.94
9. 2250 bushels of soybeans **10.** \$3.24

Objective 4.5.2

1. 150 ml of 12% solution and 300 ml of 9% solution **2.** 25 g **3.** 25 oz
4. 90 L of pure maple syrup and 30 L of 80% maple syrup **5.** 42 lb of 15% fat and 18 lb of 25% fat **6.** 30 lb
7. 66% **8.** 69.5% **9.** 200 lb of 25% polyester and 100 lb of 55% polyester **10.** 150 lb

Objective 4.6.1

1. 180 mph; 210 mph **2.** 40 mph; 60 mph **3.** 220 mi **4.** 54 mph; 42 mph **5.** $3\frac{3}{4}$ h
6. 960 mi **7.** 4 h at 55 mph; 2 h at 40 mph **8.** 60 mi **9.** $1\frac{1}{2}$ h **10.** 345 mph; 375 mph

Objective 4.7.1

1. 9 **2.** 93 or better **3.** $1020 **4.** $35,980 **5.** −11 **6.** 24 ft
7. less than 180 mi **8.** less than 840 mi **9.** $44,000 or more **10.** $27,500 or more

CHAPTER 5

Objective 5.1.1

1.
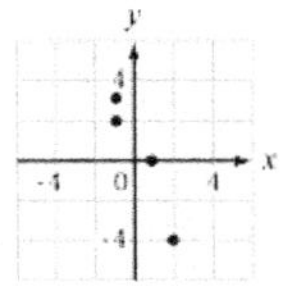

2.
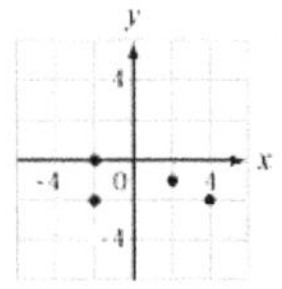

3.
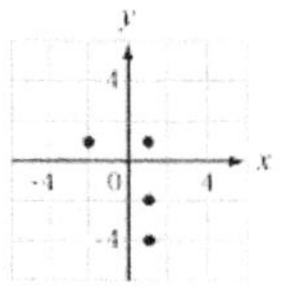

4.
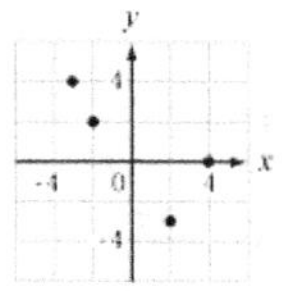

5.
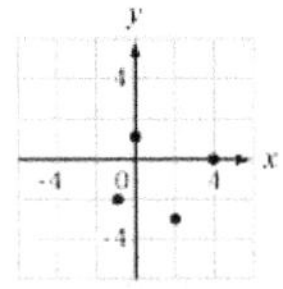

6.
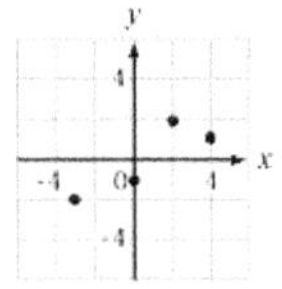

Objective 5.1.2

1.
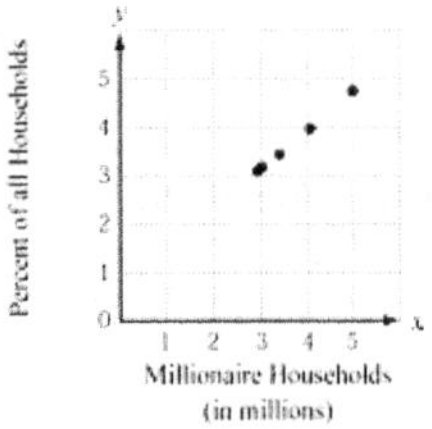

2.
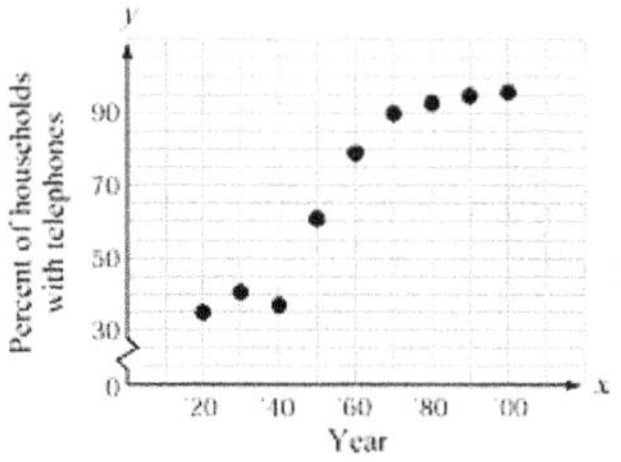

Objective 5.1.3

1. 4° per hour **2.** 0.9 cents per year **3.** 520,000 acres per year **4.** Less than **5.** 3515 lobbyists per year
6. Less than

Objective 5.2.1

1. Yes **2.** No **3.** No **4.** Yes **5.** Yes **6.** No **7.** Yes
8. No **9.** No **10.** Yes **11.** (2, 3) **12.** (−3, −7) **13.** (4, 0) **14.** (10, 1)
15. (−3, 7) **16.** (5, −3) **17.** (−4, 3) **18.** (6, −3)

Objective 5.2.2

1. 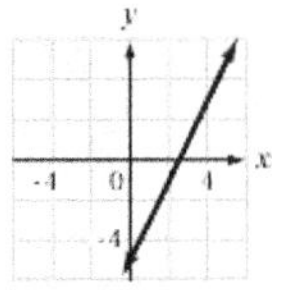
2. 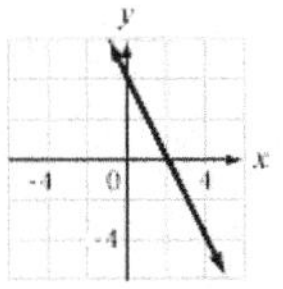
3. 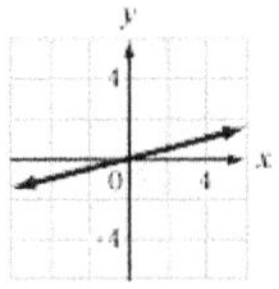
4. 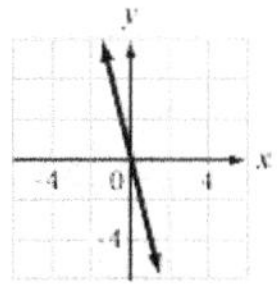
5. 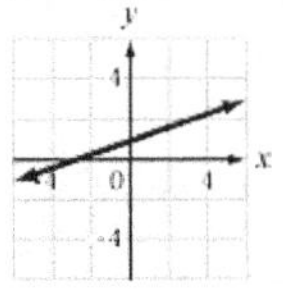
6. 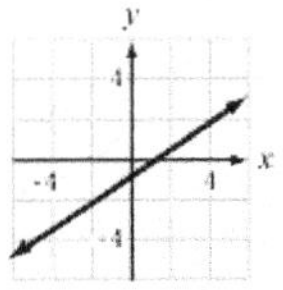
7. 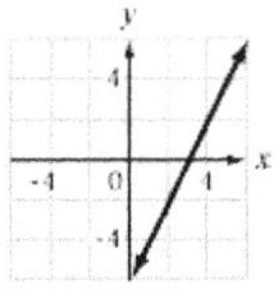
8. 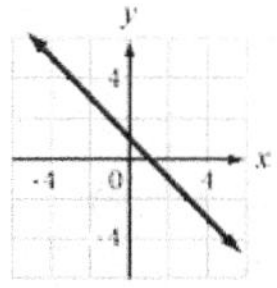

Objective 5.2.3

1. 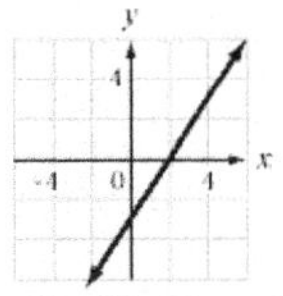
2. 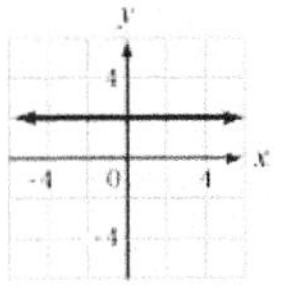
3. 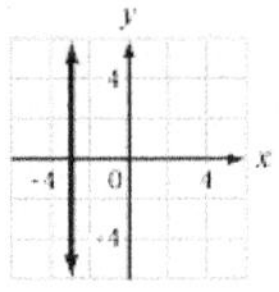
4. 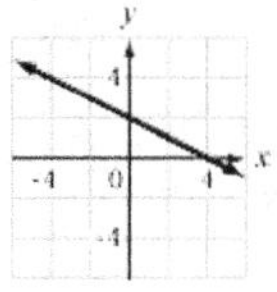

5. (5, 0); (0, 2) **6.** (–2, 0); (0, –6) **7.** (8, 0); (0, –2) **8.** (–3, 0); (0, 12) **9.** (0, 0); (0, 0) **10.** (–6, 0); (0, 2)

11. (4, 0); (0, –2)

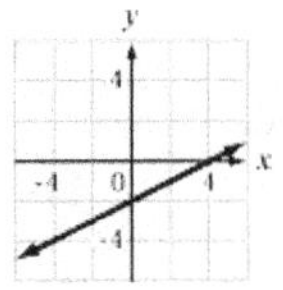

12. (–4, 0); (0, 3)

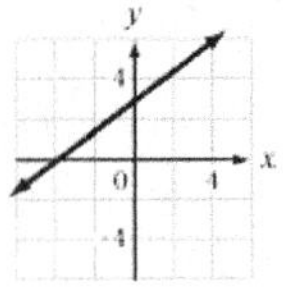

Objective 5.3.1

1. –2 **2.** 3 **3.** $\frac{1}{4}$ **4.** 0 **5.** –1 **6.** $\frac{1}{3}$ **7.** $-\frac{1}{4}$

8. –4 **9.** undefined **10.** undefined **11.** 2 **12.** $\frac{1}{5}$ **13.** 1 **14.** –2

15. 5 **16.** 3 **17.** Yes **18.** No

Objective 5.3.2

1. 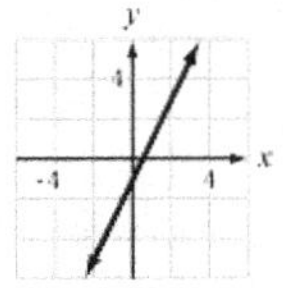
2. 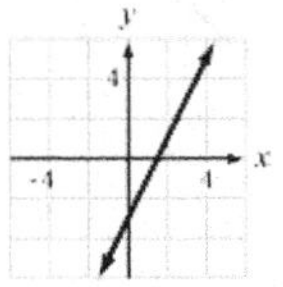
3. 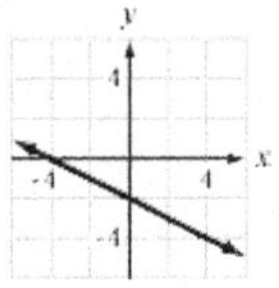
4. 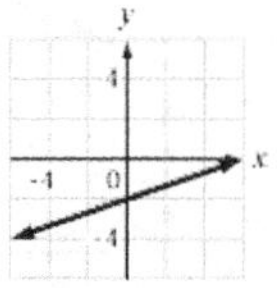
5. 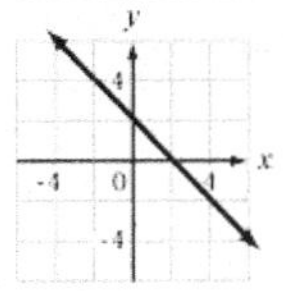
6. 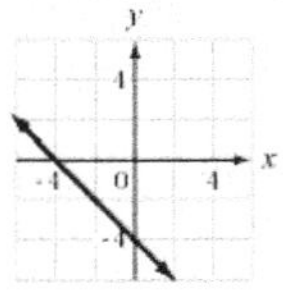
7. 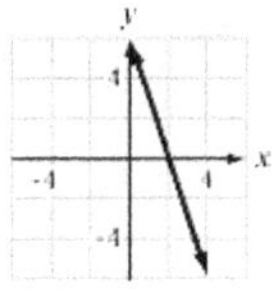
8. 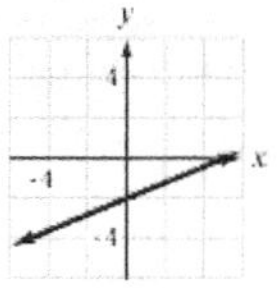

Objective 5.4.1

1. $y = x + 1$ **2.** $y = -3x - 3$ **3.** $y = -2x + 5$ **4.** $y = 2x - 4$ **5.** $y = -\frac{1}{2}x$ **6.** $y = \frac{1}{3}x + \frac{10}{3}$ **7.** $y = \frac{2}{3}x - 5$

8. $y = -\frac{1}{2}x + 5$ **9.** $y = -\frac{2}{5}x$ **10.** $y = \frac{1}{4}x - 2$ **11.** $y = -\frac{1}{3}x + \frac{8}{3}$ **12.** $y = -\frac{3}{4}x + \frac{5}{2}$

Objective 5.4.2

1. $y = x + 1$ **2.** $y = -3x - 3$ **3.** $y = -2x + 5$ **4.** $y = 2x - 4$ **5.** $y = -\frac{1}{2}x$ **6.** $y = \frac{1}{3}x + \frac{10}{3}$ **7.** $y = \frac{2}{3}x - 5$

8. $y = -\frac{1}{2}x + 5$ **9.** $y = -\frac{2}{5}x$ **10.** $y = \frac{1}{4}x - 2$ **11.** $y = -\frac{1}{3}x + \frac{8}{3}$ **12.** $y = -\frac{3}{4}x + \frac{5}{2}$

Objective 5.4.3

1. $y = -x + 5$ **2.** $y = -x + 4$ **3.** $y = \frac{2}{3}x + \frac{4}{3}$ **4.** $y = \frac{1}{2}x - \frac{1}{2}$ **5.** $y = \frac{1}{3}x + \frac{8}{3}$ **6.** $y = x$ **7.** $y = \frac{3}{2}x$

8. $y = \frac{4}{3}x + \frac{11}{3}$ **9.** $y = 2x - 2$ **10.** $y = 3x + 3$ **11.** $y = \frac{1}{2}x$ **12.** $y = -3x$ **13.** $y = -\frac{2}{3}x + \frac{2}{3}$

14. $y = -\frac{4}{3}x - \frac{4}{3}$ **15.** $y = -\frac{6}{5}x + \frac{8}{5}$ **16.** $y = -\frac{6}{7}x - \frac{11}{7}$ **17.** $y = -x + 2$ **18.** $y = -2$

Objective 5.5.1

1. Domain: {1, 2, 4, 5}; Range: {1, 3, 7}
2. Domain: {1, 3, 5, 6}; Range: {1, 2, 4}
3. Domain: {1, 2, 4, 5}; Range: {1, 3, 7}; yes
4. Domain: {–3, –1, 1, 2}; Range: {–7, –2, –1, 5, 6}; no
5. Domain: {–4, –2, 0, 3}; Range {–2, –1, 1, 4}; yes
6. Domain: {–2, –1, 3, 4}; Range: {–2, –1, 3, 4}; yes
7. –15; (–8, –15) **8.** –5; (0, –5) **9.** –2; (–1, –2) **10.** 16; (–2, 16)
11. 3; (–6, 3) **12.** 25; (–3, 25) **13.** –1; (3, –1) **14.** 43; (10, 43) **15.** 3; (0, 3)
16. Domain: {–1, 1, 3}; Range: {–1, 7, 15}
17. Domain: {–1, 0, 3}; Range: {–9, –2, 19}
18. Domain: {–2, 1, 4}; Range: {11, –1, 5}
19. Domain: {–2, 0, 4}; Range: {–4, –1, 5}

Objective 5.5.2

1.
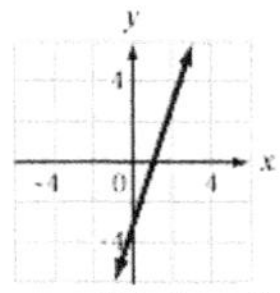

2.
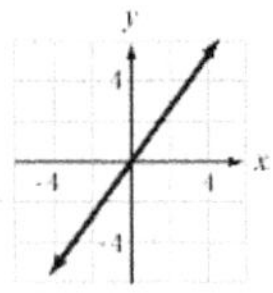

3.
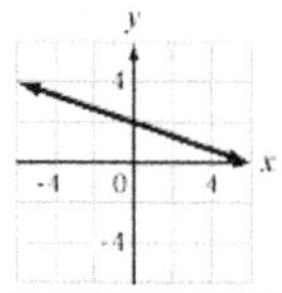

4.
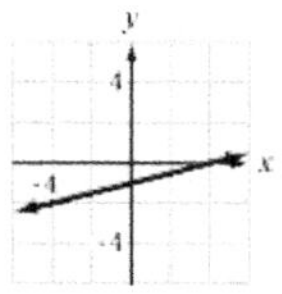

5. **a.** $C(n) = 0.25n + 7.00$

b.
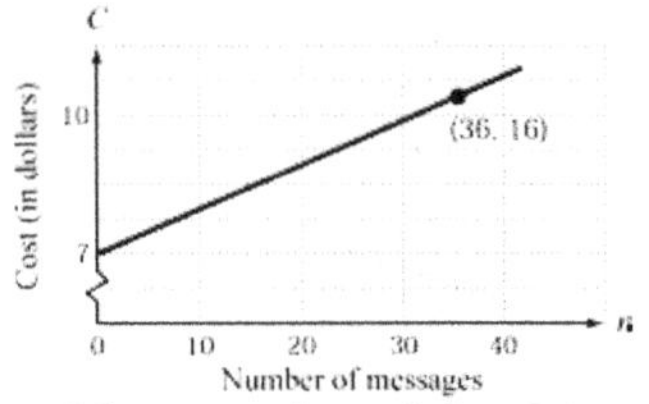

c. The cost of receiving 36 messages is \$16.

6. **a.** $C(n) = 75n + 4500$

b.
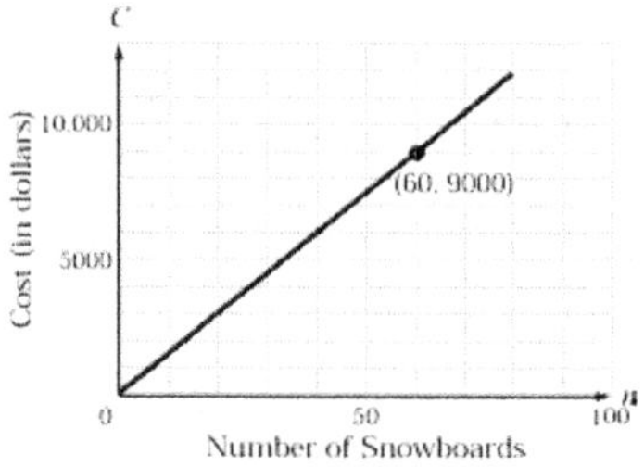

c. The cost of manufacturing 60 snow boards is \$9000.

Objective 5.6.1

1. 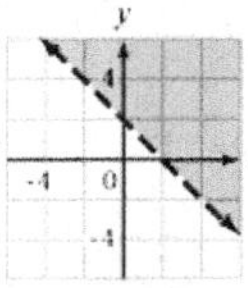
2. 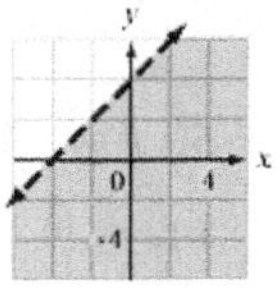
3. 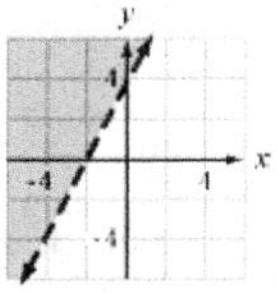
4. 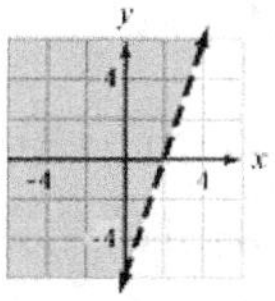
5. 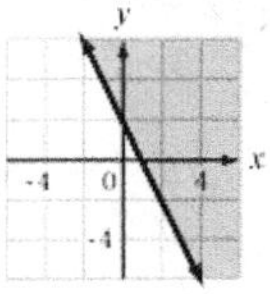
6. 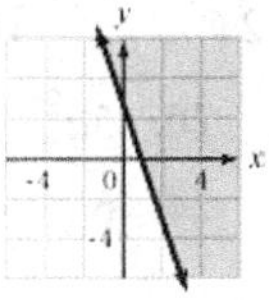
7. 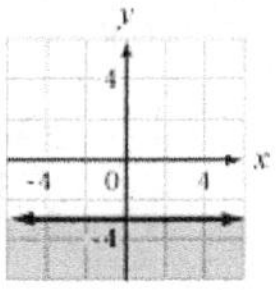
8. 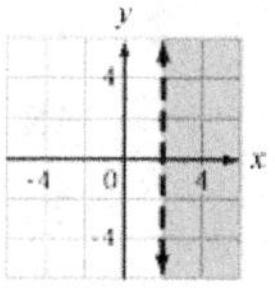

CHAPTER 6

Objective 6.1.1

1. Yes **2.** No **3.** Yes **4.** No **5.** Yes **6.** Yes **7.** dependent
8. independent **9.** inconsistent
10. The solution is (4, 2).
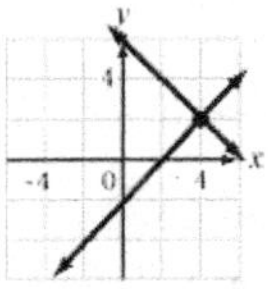
11. The solution is (–2, 5).
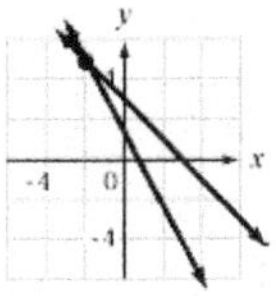
12. The solution is (3, 2).
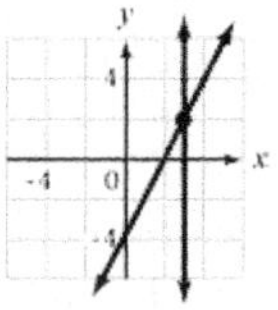
13. The solution is (2, –1).
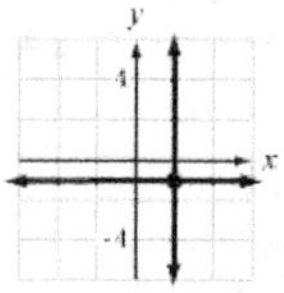

Objective 6.2.1

1. (1, 1) **2.** (3, 2) **3.** (4, 2) **4.** (1, 4) **5.** (1, 2) **6.** (5, 3) **7.** (–3, 4)
8. (4, 1) **9.** (2, –1) **10.** (1, 3) **11.** (1, –1) **12.** (1, –2) **13.** (–3, –1) **14.** (2, 7)
15. (3, –5) **16.** (2, 2) **17.** (7, 5) **18.** (–1, –1)

Objective 6.3.1

1. (6, 3) **2.** (2, 1) **3.** (–3, –2) **4.** (–2, –1) **5.** (4, 0) **6.** (5, 2) **7.** (–4, 3)
8. (–2, –5) **9.** (7, –3) **10.** (–3, 4) **11.** (1, 2) **12.** (1, –2) **13.** (3, –4) **14.** (–2, –1)
15. (3, –4) **16.** (2, 5) **17.** (3, –1) **18.** (6, –1)

Objective 6.4.1

1. rate of canoeist in calm water: 10 mph; rate of current: 2mph
2. rate of plane in calm air: 135 mph; rate of wind: 15 mph
3. rate of motorboat in still water: 12 mph; rate of current: 3 mph
4. rate of plane in calm air: 125 mph; rate of wind 25 mph
5. rate of rowing team in calm water: 7.5 mph; rate of current: 1.5 mph **6.** rate of plane: 225 mph; rate of wind 45 mph
7. rate of motorboat in calm water: 16 mph; rate of current: 4 mph
8. rate of plane in calm air: 550 mph; rate of wind: 50 mph **9.** rate of crew in calm water: 10 km/h; rate of current: 2 km/h
10. rate of plane in calm air: 350 mph; rate of wind: 50 mph

Objective 6.4.2

1. reserved-seat tickets: $9; general-admission seat tickets: $5 **2.** 4-h cassettes: $8; 8-h cassettes: $10
3. 20 quarters; 15 dimes **4.** orange tree: $6; lemon tree: $8 **5.** cinnamon tea: $2.50; spice tea: $3
6. tin: $4; zinc: $8 **7.** hamburger: $1.20; steak: $3 **8.** 30 nickels; 20 quarters

CHAPTER 7

Objective 7.1.1

1. $-x^2+2x$ **2.** $5y^2+3y$ **3.** y^2+y-5 **4.** $2x^2+7x-12$ **5.** $4x^2+10x-8$ **6.** y^2-5y-6
7. $3x^2+8x+11$ **8.** $4x^2+3x+3$ **9.** $-2y^2-9$ **10.** y^3+2y^2-2y-3 **11.** $3a^3-10a^2+7a$
12. $2y^3+2y^2-6y-2$ **13.** $5x^2+6x$ **14.** y^2+8y **15.** $7x^2+xy-2y^2$ **16.** $7x^2+4xy-3y^2$
17. $-2x^2-9x-7$ **18.** $4a^2-3a+15$ **19.** $-5x^2+5x+1$ **20.** $5a^2-6a+11$
21. $-5x^2+2x-1$ **22.** $4x^3+9x^2-x-2$ **23.** $5y^3+2y^2-8y+9$ **24.** $6x^3+x^2-x-8$

Objective 7.1.2

1. $4x$ **2.** $-6y$ **3.** $3y^2-3y-6$ **4.** $-9x-5$ **5.** $-3x^2+9x-7$ **6.** $-5x-2$ **7.** x^2-2x-9
8. $-x^2+12x+6$ **9.** $3x^3+2x^2+3x+1$ **10.** $-2x^3-4x^2-2x$ **11.** $4y^3+2y^2+4y+2$ **12.** x^3-4x-1
13. $-3xy$ **14.** $3x^2-5xy$ **15.** y^2-2xy **16.** $9y^2+y+7$ **17.** $2x^2-3x+4$ **18.** $-3x^2+7x+1$
19. $-x^3+2x^2+7$ **20.** $3x^3-2x^2-8x-5$ **21.** $2x^3+x^2+3x+6$ **22.** $3x^3-8x^2+10x-3$
23. $-3x^2+4x+1$ **24.** $-2a^3+a^2-a+6$

Objective 7.2.1

1. $3x^2$ **2.** $-2y^2$ **3.** $6x^2$ **4.** $16y^5$ **5.** $8a^5$ **6.** $-8a^{10}$ **7.** x^3y^5
8. $-12x^5y$ **9.** $-6a^6b^2$ **10.** a^3b^5 **11.** $-8x^3y^5$ **12.** $5a^3b^5$ **13.** x^3y^5z **14.** $a^3b^3c^3$
15. x^4y^4z **16.** $-a^4b^7$ **17.** $-x^4y^4$ **18.** $28y^7z^4$ **19.** $-6x^4y^2$ **20.** $a^4b^4c^6$ **21.** $x^5y^8z^3$
22. $24a^6b^6$ **23.** $-24x^5y^7$ **24.** $-6a^7b^6$ **25.** $8x^5y^8$ **26.** $-6a^{10}b^4$ **27.** $-30a^4b^3c^3$

Objective 7.2.2

1. x^4 **2.** x^6 **3.** y^8 **4.** x^{10} **5.** y^{12} **6.** x^6 **7.** $8x^3$
8. $9y^2$ **9.** $-8x^3$ **10.** $9y^4$ **11.** x^2y^4 **12.** x^8y^{12} **13.** $-8a^3b^6$ **14.** $x^{10}y^{10}$
15. x^5y^{15} **16.** $x^{16}y^{12}$ **17.** $4x^4y^5$ **18.** $-a^6b^{12}$ **19.** a^6b^6 **20.** $-54y^7$ **21.** a^4b^3
22. $-12x^5y^4$ **23.** $-18x^4y^3$ **24.** $4a^3b^8$ **25.** $-54a^5b^6$ **26.** $-24a^6b^5$ **27.** $-243a^5b^8$

Objective 7.3.1

1. x^2+x **2.** $2y-y^2$ **3.** $-x^2-2x$ **4.** $2a^2-2a$ **5.** $3b^2+15b$ **6.** $-2x^3+2x^2$ **7.** $-4y^3-24y^2$
8. $-2x^4+3x^2$ **9.** $10x^3-4x^2$ **10.** $6y^2-3y^3$ **11.** $2x^2-3x$ **12.** $6x^2-3x$ **13.** $-x^3y+x^2y^3$
14. $-2x^2y^2+xy^3$ **15.** x^3-2x^2+x **16.** x^3+3x^2-2x **17.** $-a^3+6a^2+a$
18. $-2b^3-3b^2+6b$ **19.** $2x^4-3x^3-2x^2$ **20.** $-2y^5-3y^4+4y^3$ **21.** $-4y^4-10y^3+16y^2$
22. $12x^4-6x^3+21x^2$ **23.** $20x^4-4x^3-36x^2$ **24.** $-5y^4+15y^3+30y^2$ **25.** $a^3b-3a^2b^2-4ab^3$
26. $x^3y-2x^2y^2+2xy^3$ **27.** $a^3b+5a^2b^2-7ab^3$

Objective 7.3.2

1. x^3+4x^2+5x+2 **2.** x^3-4x^2+8x-5 **3.** a^3-3a^2+4x-4 **4.** $x^3-7x^2+7x+15$
5. a^3-6a^2+8a-3 **6.** $3x^3-16x^2-7x+4$ **7.** $-2b^3+7b^2+6b-8$ **8.** $-2a^3+5a^2+7a-9$
9. $-2x^3+15x^2-24x+9$ **10.** $-3a^3-4a^2+5a+2$ **11.** x^3+x^2-4x-4 **12.** y^3-9y **13.** $x^3-5x^2+3x-15$
14. $2y^2+2y^2-4y$ **15.** $x^4-3x^3-2x^2+7x-3$ **16.** $2y^4+y^3-6y^2-10y+15$ **17.** $8y^3-14y^2-51y+7$
18. $9y^3-3y^2-11y+6$ **19.** $3a^4-6a^3-2a^2+7a-6$ **20.** $4y^3+6y^2-8y+2$ **21.** $4a^4-12a^3-4a^2+13-3$
22. $6b^4-11b^3+3b^2+18b-6$ **23.** $y^4+2y^3-y^2+2$ **24.** $3a^4-8a^3+13a^2-9a+2$

Objective 7.3.3

1. y^2+5y+6 **2.** $a^2+3a-10$ **3.** b^2-b-20 **4.** $y^2-5y-14$ **5.** $x^2+5x-36$ **6.** $y^2-8y+12$ **7.** $a^2-10a+16$
8. $x^2+8x-33$ **9.** $2x^2+13x+6$ **10.** $3y^2+5y+2$ **11.** $2x^2+3x-9$ **12.** $5x^2+13x-6$ **13.** $6x^2+13x+5$
14. $2y^2+7y-9$ **15.** $4y^2+y-14$ **16.** $30a^2+17a+2$ **17.** $12a^2-56a+65$
18. $10a^2+53a+63$ **19.** $15b^2+45b-44$ **20.** $12a^2+31a-30$ **21.** $x^2+3xy+2y^2$
22. $2a^2+5ab+2b^2$ **23.** $2x^2-5xy+3y^2$ **24.** $2a^2-3ab-9b^2$ **25.** $8a^2+18ab-5b^2$
26. $9x^2-9xy-10y^2$ **27.** $30x^2+17xy+2y^2$

Objective 7.3.4

1. x^2-1 **2.** x^2-9 **3.** x^2-49 **4.** $4x^2-1$ **5.** $9x^2-1$ **6.** $x^2+10x+25$ **7.** $y^2-8y+16$
8. $9y^2-6y+1$ **9.** x^2-2x+1 **10.** $x^2+14x+49$ **11.** $x^2+18x+81$ **12.** $x^2-2xy+y^2$
13. $4a^2-20a+25$ **14.** $25x^2-40x+16$ **15.** $9x^2-42x+49$ **16.** $9a^2-25$ **17.** $4x^2+20x+25$
18. $81x^2-36x+4$ **19.** $a^2-4ab+4b^2$ **20.** $x^2+4xy+4y^2$ **21.** $25x^2-36$ **22.** b^2-36a^2
23. $x^2+10xy+25y^2$ **24.** $4-28y+49y^2$ **25.** $9-25y^2$ **26.** $16x^2-1$ **27.** $x^2+12xy+36y^2$

Objective 7.3.5

1. $(9x^2-3x)$ m^2 **2.** $(3x^2-4x-4)$ in.^{2} **3.** $(6x^2-x-1)$ ft^2 **4.** (x^2+6x+9) cm^2
5. $(x^2-16x+64)$ km^2 **6.** $(4x^2+4x+1)$ in.^{2} **7.** $(9x^2-24x+16)$ ft^2 **8.** $(\pi x^2-6\pi x+9\pi)$ m^2
9. $(\pi x^2+12\pi x+36\pi)$ in.^{2} **10.** $(4\pi x^2+4\pi x+\pi)$ ft^2

Objective 7.4.1

1. $\frac{1}{36}$ **2.** $\frac{1}{64}$ **3.** $\frac{1}{125}$ **4.** $\frac{1}{a^5}$ **5.** $\frac{x}{y^2}$ **6.** 1 **7.** $\frac{1}{a^{14}}$
8. $\frac{y^2}{x^2}$ **9.** $\frac{1}{x^8}$ **10.** 1 **11.** $3x$ **12.** $-3x^3$ **13.** $\frac{b^6}{a^2}$ **14.** x^4y^3
15. y **16.** $5a$ **17.** $\frac{16}{a}$ **18.** $4a^2$ **19.** $\frac{-3a^2b^4}{4}$ **20.** $\frac{b^2}{-6a}$ **21.** $\frac{-2x^2y^2}{11z^3}$
22. $\frac{x^4y^3}{4}$ **23.** $-a^3b^2c$ **24.** $-\frac{4x^3}{3y^2}$

Objective 7.4.2

1. 3.56×10^{-5} **2.** 3.2×10^{5} **3.** 3.7×10^{-7} **4.** 5.1×10^{-9} **5.** 6×10^{10} **6.** 4.6×10^{7} **7.** 0.00000214
8. 0.0000000053 **9.** 6,700,000,000 **10.** 5,230,000 **11.** 0.0045 **12.** 52,400,000 **13.** 0.028
14. 0.00000001638 **15.** 0.000000004 **16.** 70,000,000 **17.** 0.000000000018 **18.** 1.3×10^{6} s
19. 4.2×10^{6} mi

Objective 7.5.1

1. $x+1$ **2.** $y+1$ **3.** $2a-3$ **4.** $2a-3$ **5.** $2b-5$ **6.** $2a+3$ **7.** $5y+3$
8. $3b^2-2$ **9.** $2y-1$ **10.** x^2+2x-4 **11.** x^2-2x-1 **12.** a^3-6a^2-4a **13.** $xy-3$
14. $-2y^2+5$ **15.** $-4x^2+3$ **16.** $-y+4$ **17.** $-2+3x^2$ **18.** $4x+1-\frac{5}{x}$ **19.** $6y+4-\frac{5}{y}$
20. $-2x+5-\frac{4}{x}$ **21.** $4a-5+6b$ **22.** $2a+1-4b$ **23.** $5x+6-2y$ **24.** $b-4+3ab$

Objective 7.5.2

1. $x+3$ **2.** $x+4$ **3.** $x-6$ **4.** $a-7$ **5.** $x-4$ **6.** $y+5$ **7.** $x+3$
8. $y-2$ **9.** $2x+3$ **10.** $2y-7$ **11.** $2y-5$ **12.** $2y-3$ **13.** $x-3$ **14.** $y+7$
15. $x-1$ **16.** $x-3+\frac{18}{x+3}$ **17.** $2x-2-\frac{6}{2x-3}$ **18.** $2y+2+\frac{2}{2y-1}$ **19.** $a-2+\frac{20}{a+5}$
20. $b-3-\frac{20}{b-4}$ **21.** $5y+3$ **22.** $6a-12+\frac{30}{2a+3}$ **23.** $2a+5$ **24.** $3x+4$

CHAPTER 8

Objective 8.1.1

1. $4(a+1)$ **2.** $6(c-1)$ **3.** $4(2-a^2)$ **4.** $3(3+7x^2)$ **5.** $3(x+3)$ **6.** $5(2a^2+3)$ **7.** $8(3a-1)$
8. $12(2x+1)$ **9.** $3(3x-2)$ **10.** $4y(3x-4)$ **11.** $b^2(6-5b)$ **12.** $4x^2(5x-6)$ **13.** $12a^2(a^3-3)$
14. $6a^2b^2(4ab^2-3)$ **15.** $2ab(2a^4+3b^3)$ **16.** $5x^2y(5y-3x)$ **17.** $xy(3x-5y)$ **18.** $4a^2b^2(3a-3b)$
19. $x(x^2-5x+7)$ **20.** $4(x^2-4x+5)$ **21.** $3(2y^2-3y+4)$ **22.** $b^2(b^2-3b+7)$
23. $4x^2(1-2x+3x^2)$ **24.** $4(3y^2-4y+12)$ **25.** $5y(y^3+2y^2-7)$ **26.** $4x^2(x^2+3x-7)$
27. $15ab(3a^2b-5a^2+2b^3)$

Objective 8.1.2

1. $(a+b)(x+3)$ **2.** $(x-y)(a+5)$ **3.** $(b-1)(x-y)$ **4.** $(a-1)(y-1)$ **5.** $(y+3)(a-1)$ **6.** $(y-7)(3x-1)$
7. $(x-5)(2x-1)$ **8.** $(x-3)(5y+1)$ **9.** $(a-2b)(x+y)$ **10.** $(x-9)(a-1)$ **11.** $(a-5)(b-2)$
12. $(x-3y)(c+d)$ **13.** $(a+1)(3x+4)$ **14.** $(a-4)(x-y)$ **15.** $(a-3b)(x-y)$ **16.** $(a-9)(x+1)$
17. $(x-7)(a-3)$ **18.** $(a-4)(2x+y)$ **19.** $(a-9)(m+n)$ **20.** $(n-2)(3m+1)$
21. $(b-1)(2a-c)$ **22.** $(a-2)(x-3y)$ **23.** $(a-b)(2-c)$ **24.** $(n-1)(m-2)$
25. $(c-7)(x+y)$ **26.** $(b-c)(a+3)$ **27.** $(x-y)(5+z)$

Answers to AIM for Success Practice Sheets

Objective 8.2.1

1. $(a-7)(a+5)$
2. $(a-3)(a-1)$
3. $(a+5)(a-1)$
4. $(b-5)(b-2)$
5. $(b+5)(b+3)$
6. $(b+8)(b-2)$
7. $(y-6)(y+5)$
8. $(y+11)(y-6)$
9. $(x-10)(x+6)$
10. $(y-2)(y-5)$
11. $(y-6)(y-3)$
12. $(x-6)(x-6)$
13. $(a+7)(a-4)$
14. $(x+8)(x+2)$
15. $(b-20)(b+9)$
16. $(x+5)(x+5)$
17. $(b+8)(b+2)$
18. $(x-12)(x+5)$
19. $(x+14)(x-4)$
20. $(x-16)(x+8)$
21. $(x-11)(x+7)$
22. $(b-14)(b-6)$
23. $(b-9)(b-12)$
24. $(b-12)(b-15)$
25. $(a+7)(a+9)$
26. $(x-15)(x-4)$
27. $(x-21)(x-4)$

Objective 8.2.2

1. $2(x+3)(x+1)$
2. $3(x+4)(x+1)$
3. $3(a+4)(a-3)$
4. $4(x-3)(x+1)$
5. $a(b+7)(b-2)$
6. $x(y-2)(y-5)$
7. $x(y+7)(y+4)$
8. $x(x-6)(x-2)$
9. $3y(y-5)(y-1)$
10. $4(x+8)(x-1)$
11. $2(x-3)(x+2)$
12. $3(z-8)(z+1)$
13. $z(z+9)(z+5)$
14. $c(c+15)(c+2)$
15. $(x-7y)(x-2y)$
16. $4(y-8)(y+4)$
17. $3(x+3)(x-2)$
18. $2(x-10)(x+3)$
19. $3(y-5)(y-3)$
20. $5(y-1)(y-9)$
21. $2x(x+3)(x-2)$
22. $(y+8x)(y-x)$
23. $x(x+8)(x-5)$
24. $10x(x+2)(x+4)$
25. $(x^2+9)(x^2-7)$
26. $2(x+15)(x-8)$
27. $a^2(a+9)(a-8)$

Objective 8.3.1

1. $(x-2)(2x-1)$
2. $(3x+1)(x-1)$
3. $(3x-2)(x+1)$
4. $(2b-1)(b-6)$
5. $(3a-1)(a-2)$
6. $(2x+3)(2x-1)$
7. $(5a-2)(a+3)$
8. $(3y-2)(2y+3)$
9. $(6x-5)(x+1)$
10. $(5x+2)(x-1)$
11. $(7y+1)(y+1)$
12. $(7x+1)(2x+1)$
13. $(7y+4)(y+2)$
14. $(3a+1)(3a-2)$
15. $(3a+4)(a-3)$
16. $(3x+2)(x-4)$
17. $(3x+2)(2x-3)$
18. $(4y+5)(y+5)$
19. $(7x-1)(x+3)$
20. $(5x+7)(x-1)$
21. $(5x+2)(2x-3)$
22. $(4x-3)(2x-5)$
23. $(4x-5)(3x+2)$
24. $(4x+5)(2x-3)$
25. $(5x+2)(2x-5)$
26. $(5x-2)(3x-4)$
27. $(6y+1)(2y-3)$

Objective 8.3.2

1. $3(x-1)^2$
2. $2(4x-1)(x+3)$
3. $2(3y-7)(y-1)$
4. $5(3y+2)(y-1)$
5. $2b(a+6)(a-2)$
6. $(4x-y)(x-6y)$
7. $(x-6y)(x-5y)$
8. $x(2x+1)(x-4)$
9. $(5+x)(3+x)$
10. $(7-x)(3+x)$
11. $(8-x)(4+x)$
12. $(2y+1)(2y-3)$
13. $3(2y-1)(2y+3)$
14. $(5x-1)(3x-4)$
15. $5(x-8)(x+1)$
16. $x(5x-1)(x-1)$
17. $(3x+4)(x-5)$
18. $3(2y-3)(2y+1)$
19. $2(2x-1)(2x+3)$
20. $5a(2a-5)(a+1)$
21. $x(3x+1)(2x+5)$
22. $2x(3x-1)(x-4)$
23. $2(5b-3)(b+2)$
24. $z(11+z)(2-z)$
25. $2x(3x-1)(x-5)$
26. $(19-x)(1+x)$
27. $(13x-3)(x+4)$

Answers to AIM for Success Practice Sheets

Objective 8.4.1

1. $(x+4)(x-4)$ **2.** $(x+8)(x-8)$ **3.** $(3x+1)(3x-1)$ **4.** $(4x+5)(4x-5)$ **5.** $(x^4+10)(x^4-10)$
6. $(9x+1)(9x-1)$ **7.** $(y^2+11)(y^2-11)$ **8.** nonfactorable **9.** $(x^2+y^4)(x^2-y^4)$ **10.** $(b+12c)(b-12c)$
11. $(x^3+9)(x^3-9)$ **12.** $(xy+1)(xy-1)$ **13.** $(x+20)(x-20)$ **14.** $(6a+1)(6a-1)$
15. $(7x+2)(7x-2)$ **16.** $(x+3)^2$ **17.** nonfactorable **18.** $(x+8)^2$ **19.** $(5-a)^2$ **20.** $(6x-1)^2$
21. $(10x-1)^2$ **22.** $(2x+9)^2$ **23.** $(2a+11)^2$ **24.** $(7a+5)^2$ **25.** $(2x-5y)^2$ **26.** $(2x+3y)^2$ **27.** $(3x-5y)^2$

Objective 8.5.1

1. $3(x+2)(x-2)$ **2.** $2(x+5)(x-5)$ **3.** $x(x+1)^2$ **4.** $y(y-4)^2$ **5.** $x^2(x+3)(x-2)$
6. $a^2(a-8)(a+5)$ **7.** $3(b+3)(b+7)$ **8.** $4(y-1)(y-7)$ **9.** $2(a-11)(a+2)$
10. $x(x-10)(x+2)$ **11.** $b(b-6)(b+1)$ **12.** $5a(a-3)^2$ **13.** $2(2x-1)(x-1)$ **14.** $x^2(2x-1)(x-5)$
15. $x^2(x+4)(x-4)$ **16.** $3(5x-1)(x-1)$ **17.** $3a(3a+1)(a+4)$ **18.** $x(3y-4)(y+5)$
19. $3(4-x)(2+x)$ **20.** $b^2(a+8)(a-1)$ **21.** $4y(x+2)(x+1)$ **22.** $18a(a-2)(a-1)$
23. $y^2(x+1)(x-1)$ **24.** $5(x+3y)(x-3y)$ **25.** $x^2(x+3)(x-3)$ **26.** $(3x-3)(a+b)$
27. $(a+b)(y+2)(y-2)$

Objective 8.6.1

1. $-2, -1$ **2.** $4, 6$ **3.** $1, 6$ **4.** $-7, 5$ **5.** $0, 8$ **6.** $-1, 0$ **7.** $0, 4$
8. $-\frac{2}{3}, 0$ **9.** $0, \frac{5}{2}$ **10.** $0, \frac{1}{2}$ **11.** $-\frac{3}{4}, 0$ **12.** $-1, 4$ **13.** $-4, 7$ **14.** $-3, 7$
15. $-8, 2$ **16.** $-1, 6$ **17.** $-2, 7$ **18.** $-1, \frac{3}{2}$ **19.** $\frac{1}{4}, 3$ **20.** $-3, \frac{2}{5}$ **21.** $-2, -\frac{1}{2}$
22. $-11, 1$ **23.** $3, 6$ **24.** $-10, 5$ **25.** $-4, \frac{7}{2}$ **26.** $-4, 12$ **27.** $-4, \frac{3}{2}$

Objective 8.6.2

1. 4 **2.** −6 **3.** 2, 5 **4.** 4, 6 **5.** 3, 9 **6.** 3 **7.** 5, 6
8. 10, 12 **9.** base: 24 ft; height: 8 ft **10.** length: 24 in.; width: 6 in. **11.** length: 15 ft; width: 8 ft **12.** 8 in.

CHAPTER 9

Objective 9.1.1

1. $\frac{2}{3x^2}$ **2.** $\frac{3x}{4y^2}$ **3.** $\frac{1}{(x+1)^2}$ **4.** $(3x-2)^2$ **5.** -1 **6.** $\frac{3}{4y}$ **7.** $\frac{1}{3y}$
8. $\frac{4x^2}{3}$ **9.** $\frac{x^2}{2}$ **10.** $\frac{a}{c}$ **11.** $\frac{x}{3}$ **12.** $-\frac{2}{x}$ **13.** $\frac{x+5}{x-7}$ **14.** $\frac{x-3}{x-6}$
15. $\frac{x-3}{x+3}$ **16.** $\frac{x+2}{x-1}$ **17.** $\frac{x-4}{x-5}$ **18.** $\frac{x-2}{x+1}$ **19.** $\frac{x-6}{4-x}$ **20.** $\frac{-1(y+3)}{y+5}$ **21.** $\frac{3(x-1)}{x-2}$
22. $\frac{1}{2(x+1)}$ **23.** $\frac{x-3}{x+5}$ **24.** $\frac{x-4}{2-x}$ **25.** $\frac{2(x+2)}{x+6}$ **26.** $\frac{2x-7}{x+3}$ **27.** $2(x-1)$

Answers to AIM for Success Practice Sheets

Objective 9.1.2

1. $\frac{2b}{25ax^2y}$ **2.** $\frac{ay}{20x}$ **3.** $\frac{15}{4}$ **4.** $\frac{3}{4}$ **5.** $\frac{x+1}{x+4}$ **6.** $\frac{x}{x+2}$ **7.** $\frac{x+2}{x(x-2)}$

8. $\frac{x(x+4)}{y(x-5)}$ **9.** $\frac{y}{2}$ **10.** $\frac{x-1}{x^2(x+8)}$ **11.** 1 **12.** $\frac{x-5}{x+5}$ **13.** $\frac{x-9}{x+9}$ **14.** $\frac{x+9}{x+2}$

15. $\frac{x+7}{x-3}$ **16.** $\frac{x+6}{x-1}$ **17.** $x+1$ **18.** $\frac{2x-3}{x+6}$ **19.** $\frac{3x-1}{2x+3}$ **20.** $\frac{x+3}{2x+3}$

Objective 9.1.3

1. $\frac{9xy^2}{20ab}$ **2.** $\frac{8b}{3}$ **3.** $\frac{1}{2}$ **4.** $\frac{5}{4}$ **5.** $\frac{x}{x+2}$ **6.** $\frac{x-3}{x+1}$ **7.** $\frac{x(x+1)}{x+2}$

8. $x+3$ **9.** $\frac{x(x+5)}{y^2(x-5)}$ **10.** $3ay^2$ **11.** 1 **12.** $\frac{x-5}{x-7}$ **13.** 1 **14.** $\frac{x-1}{x+1}$

15. $\frac{x+4}{x-4}$ **16.** $\frac{x+6}{x-6}$ **17.** $\frac{(x+4)(x+2)}{(x+3)^2}$ **18.** $\frac{x+5}{x-12}$ **19.** $\frac{1}{x-2}$ **20.** $\frac{3y-2}{3y-1}$

Objective 9.2.1

1. $18x^2y^2$ **2.** $30ab^3$ **3.** $60x^3y^3$ **4.** $60x^3y^2$ **5.** $10x^3(x+2)$ **6.** $8x^3(3x+2)$ **7.** $12xy^2(x+3)$

8. $72x(x+1)^2$ **9.** $36x^2(x-1)^2$ **10.** $2(x+3)(x+1)$ **11.** $3(x-4)(x-1)$ **12.** $(x-2)(x+3)(x+4)$

13. $(2x-3)(2x+3)(x+2)$ **14.** $(3x+1)^2(x-4)$ **15.** $(x+2)(x-3)$ **16.** $(x+5)(x-3)(x-4)$

17. $(x-6)(x+3)(x+4)$ **18.** $(x-1)(x-5)(x+7)$ **19.** $(x-6)(x-2)(x-3)$ **20.** $(x+1)(x+8)(x-8)$

21. $(x-5)(x-7)(x+6)$ **22.** $(2x-3)(x-2)(x+2)$ **23.** $(2x-1)(x-7)(x+5)$ **24.** $-1(x-6)(x+4)$

25. $-1(x-7)(x+2)$ **26.** $-1(5+x)(3-x)$ **27.** $-(x-6)(x-2)$

Objective 9.2.2

1. $\frac{2}{x^2}, \frac{5x}{x^2}$ **2.** $\frac{4b}{a^2b^2}, \frac{7a}{a^2b^2}$ **3.** $\frac{5xy}{10y^2}, \frac{2z}{10y^2}$ **4.** $\frac{6x^2}{12x^2y^2}, \frac{15y}{12x^2y^2}$ **5.** $\frac{x(y-1)}{y(y-1)}, \frac{5}{y(y-1)}$

6. $\frac{a(y+2)}{y(y+2)}, \frac{3}{y(y+2)}$ **7.** $\frac{5x^2}{x^2(x-2)}, \frac{7}{x^2(x-2)}$ **8.** $\frac{a(y+6)}{y(y+6)^2}, \frac{a^2y}{y(y+6)^2}$ **9.** $\frac{-4x}{x(2-x)}, \frac{7}{x(2-x)}$

10. $\frac{4}{(x-1)^2}, \frac{2(x-1)}{(x-1)^2}$ **11.** $\frac{-6(y-6)}{(y-6)^2}, \frac{3}{(y-6)^2}$ **12.** $\frac{a}{x(x-3)}, \frac{-a^2x}{x(x-3)}$ **13.** $\frac{2}{(x-6)^2}, \frac{-3(x-6)}{(x-6)^2}$

14. $\frac{5x}{x^2(x+3)}, \frac{6(x+3)}{x^2(x+3)}$ **15.** $\frac{3x}{x(x-2)}, \frac{4}{x(x-2)}$ **16.** $\frac{(x-1)(x-3)}{(x+2)(x-3)}, \frac{x(x+2)}{(x+2)(x-3)}$ **17.** $\frac{x(x+3)}{(x-7)(x+3)}, \frac{(x+2)(x-7)}{(x-7)(x+3)}$

18. $\frac{2}{(x-3)(x+1)}, \frac{x(x+1)}{(x-3)(x+1)}$ **19.** $\frac{2x(x+2)}{(x-2)(x+2)}, \frac{3}{(x+2)(x-2)}$ **20.** $\frac{4(x-4)}{(x+4)(x-4)^2}, \frac{(x+1)(x+4)}{(x+4)(x-4)^2}$

21. $\frac{x-4}{(2x-3)(x+3)}, \frac{3(2x-3)}{(2x-3)(x+3)}$ **22.** $\frac{x-1}{(3x-1)(x+2)}, \frac{3(3x-1)}{(3x-1)(x+2)}$

23. $\frac{x(x+1)}{(x+2)(x-1)(x+1)}, \frac{3x(x+2)}{(x+2)(x-1)(x+1)}$ **24.** $\frac{(x-2)(x-3)}{(x+4)(x-1)(x-3)}, \frac{x(x+4)}{(x+4)(x-1)(x-3)}$

25. $\frac{-x(x-2)}{(x+2)(x-2)^2}, \frac{(x-3)(x+2)}{(x+2)(x-2)^2}$ **26.** $\frac{x(x+3)}{(x-1)(x+3)}, \frac{2x(x-1)}{(x-1)(x+3)}, \frac{-2}{(x-1)(x+3)}$

27. $\frac{(x+2)(x-5)}{(x+4)(x-5)}, \frac{(x+1)(x+4)}{(x+4)(x-5)}, \frac{-2}{(x+4)(x-5)}$

Objective 9.3.1

1. $\frac{7}{x^2}$ 2. $\frac{6}{xy}$ 3. $\frac{-5}{x+3}$ 4. $\frac{x-3}{x-5}$ 5. $\frac{10x}{3x+2}$ 6. $\frac{-3y}{5y-1}$ 7. $\frac{2(2x+3)}{x-2}$

8. $\frac{8x-5}{2x-1}$ 9. 2 10. 3 11. $\frac{-2(2x+1)}{3x+5}$ 12. $\frac{-3n+5}{2n+9}$ 13. $\frac{1}{x-5}$ 14. $\frac{2}{x+3}$

15. $\frac{1}{x-2}$ 16. $\frac{1}{x-2}$ 17. $\frac{1}{y+5}$ 18. $\frac{5}{2x+3}$ 19. $\frac{4}{x-5}$ 20. $\frac{x+2}{x+7}$

Objective 9.3.2

1. $\frac{20}{3x}$ 2. $\frac{5}{6a}$ 3. $\frac{17}{18x}$ 4. $\frac{75+y}{15y^2}$ 5. $\frac{8x-1}{6x}$ 6. $\frac{11x+2}{24x}$ 7. $\frac{7x+55}{36x}$

8. $\frac{x+1-3xy+3y}{xy^2}$ 9. $\frac{3x^2-5x-9}{6x^2}$ 10. $\frac{4x^2+13x+12}{36x^2}$ 11. $\frac{x+11}{(x-1)(x+2)}$ 12. $\frac{3(y-9)}{(y+6)(y-3)}$

13. $\frac{2(x^2+2x-32)}{(x-8)(x-2)}$ 14. $\frac{3(x^2-7x+2)}{(2x-1)(x-3)}$ 15. $\frac{-x-4}{(x+2)(x-2)}$ 16. $\frac{7x+6}{(x-4)(x+2)}$ 17. $\frac{4}{x-3}$

18. $\frac{x^2+5x-5}{(x-2)(x-1)}$

Objective 9.4.1

1. $\frac{x}{x+5}$ 2. $\frac{2x(x+3)}{(x+6)(x-6)}$ 3. $\frac{1}{2}$ 4. 3 5. $\frac{5(2x-1)}{3(2x-5)}$ 6. $\frac{2}{3}$ 7. $\frac{x-9}{x+3}$

8. $\frac{x+2}{x-7}$ 9. $\frac{x-6}{x+3}$ 10. $\frac{x-2}{x+3}$ 11. $\frac{x}{x+6}$ 12. $\frac{x^2-x+1}{x}$ 13. $\frac{x^2+5x+18}{x^2+4x+1}$

14. $\frac{x+3}{x+4}$ 15. $\frac{x+2}{7x+6}$ 16. $\frac{-2}{3(3x+2)}$

Objective 9.5.1

1. 2 2. 3 3. 3 4. 1 5. $\frac{4}{3}$ 6. 4 7. 3

8. 7 9. 3 10. 2 11. –4 12. $\frac{1}{2}$ 13. 4 14. 5

15. 6 16. –4 and 2 17. 2 and 3 18. –6 and 4 19. $\frac{3}{2}$ and 2 20. 1 and 12

Objective 9.5.2

1. 15 2. 30 3. 25 4. 16 5. 12 6. 24 7. 40
8. 6 9. 17 10. 28 11. 17 12. 18 13. 64 14. 1

15. 27 16. 34 17. $-\frac{1}{2}$ 18. $-\frac{8}{3}$ 19. 7 20. –1 21. $-\frac{3}{2}$

22. 21 23. $\frac{7}{2}$ 24. 2 25. $-\frac{3}{5}$ 26. 13 27. $-\frac{1}{9}$

Objective 9.5.3

1. \$217.35 **2.** \$1500 **3.** 216 sq ft **4.** 16,000 voters **5.** 90 clocks **6.** \$96 **7.** 250 deer
8. \$50 **9.** 1 oz **10.** 3 gal

Objective 9.5.4

1. 6.9 cm **2.** 15.8 in. **3.** 3.7 m **4.** 28.5 ft **5.** 16.2 m **6.** 108 m^2 **7.** 300 $in.^2$
8. 7.2 cm **9.** 9 in. **10.** 13.75 ft

Objective 9.6.1

1. $n=\frac{a_n-a_1}{d}+1$ **2.** $b=\frac{2A}{h}$ **3.** $a=-\frac{b}{2a}$ **4.** $n=\frac{PV}{RT}$ **5.** $G=\frac{Fr^2}{m_1m_2}$ **6.** $V_2=\frac{P_1V_1T_2}{P_2T_1}$
7. $b_1=\frac{2A-hb_2}{h}$ **8.** $r=\frac{D}{t}$ **9.** $C=\frac{100M}{I}$ **10.** $T=\frac{E+alt}{al}$ **11.** $a=\frac{bt}{b-t}$ **12.** $v=\frac{2at-2d}{a}$
13. $g=\frac{2h-2vt}{t^2}$ **14.** $C=Rt+S$ **15.** $m=\frac{RK+T}{B}$ **16.** $m=\frac{K}{r-v}$ **17.** $I=\frac{\pi r^2-S}{\pi r}$ **18.** $c=\frac{A-P}{A}$

Objective 9.7.1

1. $1\frac{1}{2}$ h **2.** 6 min **3.** 15 min **4.** 24 min **5.** 12 min **6.** 24 h **7.** $20\frac{4}{5}$ h
8. 21 h **9.** 8 h **10.** 32 h

Objective 9.7.2

1. freight train: 35 mph; express train: 55 mph **2.** commuter plane: 180 mph; jet plane: 380 mph
3. car: 50 mph; plane: 400 mph **4.** single-engine plane: 97.5 mph; jet plane: 390 mph **5.** 7 mph
6. 50 mph **7.** 27 mph **8.** 30 mph **9.** 2 mph **10.** 40 mph

CHAPTER 10

Objective 10.1.1

1. 5 **2.** 11 **3.** 13 **4.** 14 **5.** 17 **6.** 20 **7.** $4\sqrt{3}$
8. $2\sqrt{5}$ **9.** $3\sqrt{5}$ **10.** $6\sqrt{2}$ **11.** $-24\sqrt{2}$ **12.** $18\sqrt{10}$ **13.** $5\sqrt{5}$ **14.** $-35\sqrt{2}$
15. $8\sqrt{3}$ **16.** 36 **17.** $5\sqrt{6}$ **18.** 30 **19.** $10\sqrt{2}$ **20.** $12\sqrt{10}$ **21.** $35\sqrt{2}$
22. $2\sqrt{22}$ **23.** $3\sqrt{15}$ **24.** $8\sqrt{2}$ **25.** $10\sqrt{7}$ **26.** $7\sqrt{5}$ **27.** $8\sqrt{6}$

Objective 10.1.2

1. x^4 **2.** $y^4\sqrt{y}$ **3.** $x^6\sqrt{x}$ **4.** x^3y^3 **5.** x^5y^2 **6.** $3x^3$ **7.** $4y^5$
8. $xy^4\sqrt{xy}$ **9.** $a^6b\sqrt{ab}$ **10.** $4y^2\sqrt{2y}$ **11.** $3x^2\sqrt{3x}$ **12.** $4x^4y^2\sqrt{3xy}$ **13.** $4xy^2\sqrt{6}$ **14.** $6ab^3$
15. $10x^2y^8$ **16.** $3x^4y^2\sqrt{10xy}$ **17.** $16a^3b\sqrt{b}$ **18.** $27ab^2\sqrt{b}$ **19.** x^2y^2 **20.** $x^2y^5\sqrt{x}$
21. $9ab\sqrt{a}$ **22.** $24a^3b^4\sqrt{b}$ **23.** $16xy^3\sqrt{2y}$ **24.** $8(x+y)^2$ **25.** $7(x-3)$ **26.** $2(x-1)$ **27.** $b-5$

Objective 10.2.1

1. $5\sqrt{3}$ **2.** $11\sqrt{2}$ **3.** $\sqrt{6}$ **4.** $-9\sqrt{2}$ **5.** $-10\sqrt{5}$ **6.** $10\sqrt{y}$ **7.** $-4\sqrt{x}$
8. $-\sqrt{3a}$ **9.** $-12\sqrt{7b}$ **10.** $-10\sqrt{2a}$ **11.** $3x\sqrt{3}$ **12.** $-3a\sqrt{5a}$ **13.** $-7x\sqrt{3y}$ **14.** $-8\sqrt{xy}$
15. $2\sqrt{2y}$ **16.** $4\sqrt{2}$ **17.** $3\sqrt{5}$ **18.** $\sqrt{x}$ **19.** $36\sqrt{y}$ **20.** $2\sqrt{x}$ **21.** $24\sqrt{y}$
22. $-10y\sqrt{2}$ **23.** $4\sqrt{2}$ **24.** $-\sqrt{5}$ **25.** $-\sqrt{5x}$ **26.** $19\sqrt{3}$ **27.** $25\sqrt{2}$

Objective 10.3.1

1. 9 **2.** 10 **3.** z **4.** x^2y^3 **5.** a^3b^2 **6.** $4a^2b^3$ **7.** $6x^3y^2$
8. $10a^4b^2$ **9.** $5a^5b^4\sqrt{3}$ **10.** $6-3\sqrt{2}$ **11.** $4\sqrt{2}$ **12.** -2 **13.** $x+\sqrt{xy}$ **14.** $6-\sqrt{3x}$
15. $2\sqrt{2y}-4$ **16.** $2\sqrt{3}-6$ **17.** $10-\sqrt{5a}$ **18.** $x-8\sqrt{x}+16$ **19.** $9x-6y\sqrt{x}+y^2$ **20.** $2a-4\sqrt{ab}$
21. $3x\sqrt{2}-3x$ **22.** $6x-19y\sqrt{x}+15y^2$ **23.** $20x+\sqrt{xy}-y$ **24.** $4x-y$ **25.** $5x-y^2$
26. $12x-5\sqrt{xy}-3y$ **27.** $8x-26\sqrt{xy}+15y$

Objective 10.3.2

1. 5 **2.** 3 **3.** 4 **4.** 9 **5.** 6 **6.** $5x$ **7.** $2y$
8. $\frac{\sqrt{5}}{5}$ **9.** $\frac{\sqrt{2}}{2}$ **10.** $\frac{5\sqrt{x}}{x}$ **11.** $\frac{2\sqrt{3x}}{x}$ **12.** $\frac{4x}{y}$ **13.** $\frac{y^2\sqrt{y}}{3x}$ **14.** $\frac{4y^2}{x}$
15. $\frac{\sqrt{3}}{xy}$ **16.** $\frac{\sqrt{3}+1}{2}$ **17.** $3(\sqrt{5}+2)$ **18.** $3(3-\sqrt{6})$ **19.** $\frac{6(\sqrt{2}+5)}{-23}$ **20.** $\frac{\sqrt{3}+3}{-3}$ **21.** $\frac{x-\sqrt{xy}}{x-y}$
22. -3 **23.** 2 **24.** 5 **25.** $2x-6$ **26.** $9y^2$ **27.** $3a^2$

Objective 10.4.1

1. 64 **2.** 256 **3.** 100 **4.** 3 **5.** 8 **6.** 12 **7.** 5
8. 24 **9.** $\frac{9}{2}$ **10.** 14 **11.** -1 **12.** 1 **13.** 5 **14.** 5
15. 4 **16.** 8 **17.** 7 **18.** 20 **19.** 8 **20.** 2 **21.** 24
22. $-\frac{5}{2}$ **23.** $\frac{2}{3}$ **24.** 29 **25.** no solution **26.** 2 **27.** 1

Objective 10.4.2

1. 8 **2.** 2 **3.** 8.16 ft **4.** 9 ft **5.** 144 ft **6.** 196 ft **7.** 64 ft
8. 256 ft **9.** 12.98 ft **10.** 4.67 ft

CHAPTER 11

Objective 11.1.1

1. 4, 6 **2.** -3, 6 **3.** 1, 5 **4.** 1, 6 **5.** -8, -1 **6.** -9, -1 **7.** 5
8. 6 **9.** $-\frac{1}{2}, 0$ **10.** $0, \frac{3}{2}$ **11.** -3, 7 **12.** -4, 5 **13.** $-\frac{5}{2}, 2$ **14.** $-\frac{3}{2}$
15. $-\frac{4}{5}, 3$ **16.** $-\frac{2}{3}, 1$ **17.** $\frac{1}{5}, 2$ **18.** $\frac{1}{4}$ **19.** $\frac{1}{5}$ **20.** -1, 1 **21.** -5, 5
22. $-\frac{1}{4}, \frac{1}{4}$ **23.** $-\frac{5}{2}, \frac{5}{2}$ **24.** -4, 7 **25.** -2, 6 **26.** $-\frac{1}{2}, 1$ **27.** -3, -1

Answers to AIM for Success Practice Sheets

Objective 11.1.2

1. –4, 4 **2.** –9, 9 **3.** –12, 12 **4.** $-\frac{5}{3}, \frac{5}{3}$ **5.** $-\frac{9}{4}, \frac{9}{4}$ **6.** $-\frac{5}{4}, \frac{5}{4}$ **7.** $-\frac{9}{2}, \frac{9}{2}$

8. $-\frac{4}{5}, \frac{4}{5}$ **9.** $-\frac{7}{6}, \frac{7}{6}$ **10.** $-\frac{3}{2}, \frac{3}{2}$ **11.** $-\frac{10}{3}, \frac{10}{3}$ **12.** No real number solution

13. No real number solution **14.** $-3\sqrt{2}, 3\sqrt{2}$ **15.** $-5\sqrt{2}, 5\sqrt{2}$ **16.** –4, 2 **17.** –4, 8

18. –8, –2 **19.** –2, 6 **20.** $-\frac{3}{2}, \frac{7}{2}$ **21.** $-\frac{16}{3}, \frac{4}{3}$ **22.** $-\frac{19}{4}, -\frac{5}{4}$ **23.** $-\frac{1}{5}, \frac{11}{5}$

24. $3-4\sqrt{2}, 3+4\sqrt{2}$ **25.** $-4-5\sqrt{3}, -4+5\sqrt{3}$ **26.** $1+5\sqrt{2}, 1-5\sqrt{2}$ **27.** $-1-4\sqrt{5}, -1+4\sqrt{5}$

Objective 11.2.1

1. –7, 1 **2.** –7, 3 **3.** –3, 9 **4.** –7, 5 **5.** 3 **6.** –3, 1

7. $2-\sqrt{11}, 2+\sqrt{11}$ **8.** $-4-\sqrt{23}, -4+\sqrt{23}$ **9.** $3-\sqrt{10}, 3+\sqrt{10}$ **10.** $2-\sqrt{7}, 2+\sqrt{7}$ **11.** 4

12. –3, 1 **13.** $\frac{-3-\sqrt{21}}{2}, \frac{-3+\sqrt{21}}{2}$ **14.** 1, 3 **15.** $-4, \frac{3}{2}$ **16.** $-\frac{1}{2}, 3$ **17.** $-5, \frac{1}{2}$

18. $\frac{1}{3}, 2$ **19.** $-\frac{2}{3}, 1$ **20.** $-2, \frac{1}{4}$ **21.** $-\frac{2}{3}, 2$ **22.** $-1, \frac{5}{3}$ **23.** $-\frac{1}{4}, 1$ **24.** $-\frac{1}{3}, 1$

25. $\frac{1}{3}, \frac{1}{2}$ **26.** $-\frac{1}{2}, \frac{5}{2}$ **27.** $-\frac{1}{2}, \frac{2}{3}$

Objective 11.3.1

1. –1, 4 **2.** –5, –1 **3.** –2, 5 **4.** 1, 5 **5.** –6, 1 **6.** –4, 5 **7.** –2, 3

8. –1, 9 **9.** –2, 8 **10.** –9, 1 **11.** –10, 1 **12.** $-2, \frac{5}{2}$ **13.** $3-\sqrt{3}, 3+\sqrt{3}$

14. $-3-\sqrt{7}, -3+\sqrt{7}$ **15.** 0, 4 **16.** $0, \frac{1}{3}$ **17.** $-\frac{5}{3}, \frac{5}{3}$ **18.** $-\frac{7}{4}, \frac{7}{4}$ **19.** $-\frac{7}{3}, 2$

20. $-\frac{1}{2}, 4$ **21.** $-2, \frac{4}{3}$ **22.** $-\frac{5}{2}, 2$ **23.** $-\frac{2}{5}, 3$ **24.** $-1, \frac{2}{3}$ **25.** $\frac{1-\sqrt{13}}{2}, \frac{1+\sqrt{13}}{2}$

26. $-3-\sqrt{13}, -3+\sqrt{13}$ **27.** $\frac{-5-\sqrt{29}}{2}, \frac{-5+\sqrt{29}}{2}$

Objective 11.4.1

1. $-9i\sqrt{3}$ **2.** $2i\sqrt{2}$ **3.** $2i\sqrt{3}$ **4.** $10i\sqrt{2}$ **5.** $8i\sqrt{2}$ **6.** $12i$ **7.** $6i\sqrt{3}$

8. $6i\sqrt{5}$ **9.** $3+3i$ **10.** $4+5i$ **11.** $6+4i$ **12.** $7+10i$ **13.** $4\sqrt{3}-3i\sqrt{3}$ **14.** $2\sqrt{6}-6i$

15. $5\sqrt{2}-7i\sqrt{2}$ **16.** $5\sqrt{6}-3i\sqrt{3}$

Objective 11.4.2

1. $-2i$ **2.** $21i$ **3.** $8i$ **4.** $11-6i$ **5.** $11-9i$ **6.** $12+3i$ **7.** $12-i$

8. $13-7i$ **9.** $9-i$ **10.** $8-3i$ **11.** $6-i$ **12.** $7-i$ **13.** 0 **14.** $-3-3i$

15. $2-i$ **16.** $5+2i$ **17.** $-7+17i$ **18.** $-8+5i$

Objective 11.4.3

1. 40 **2.** –15 **3.** $-160i$ **4.** –32 **5.** $-3+15i$ **6.** $-12+8i$ **7.** $17-i$

8. $9+13i$ **9.** $9-7i$ **10.** $2-11i$ **11.** $13-i$ **12.** $3-21i$ **13.** $6+13i$ **14.** $24-10i$

15. $21+20i$ **16.** 40 **17.** 34 **18.** 5

Objective 11.4.4

1. $-4i$ **2.** $-\frac{2}{3}i$ **3.** $-\frac{3}{4}i$ **4.** $2i+3$ **5.** $3i-\frac{2}{5}$ **6.** $\frac{20}{17}+\frac{5}{17}i$ **7.** $\frac{12}{5}-\frac{4}{5}i$
8. $\frac{16}{17}+\frac{4}{17}i$ **9.** $\frac{15}{13}+\frac{3}{13}i$ **10.** $-i$ **11.** $-i$ **12.** $\frac{20}{13}+\frac{48}{13}i$ **13.** $\frac{21}{10}-\frac{3}{10}i$ **14.** $4i$
15. $\frac{4}{5}-\frac{7}{5}i$ **16.** $1+2i$ **17.** $-\frac{3}{2}+\frac{7}{2}i$ **18.** $\frac{7}{5}+\frac{16}{5}i$

Objective 11.4.5

1. $3i, -3i$ **2.** $8i, -8i$ **3.** $3i, -3i$ **4.** $7i, -7i$ **5.** $3+2i, 3-2i$ **6.** $-2+3i, -2-3i$
7. $-1+i\sqrt{2}, -1-i\sqrt{2}$ **8.** $-3+4i, -3-4i$ **9.** $1+i\sqrt{5}, 1-i\sqrt{5}$ **10.** $-4+i, -4-i$
11. $-1+6i, -1-6i$ **12.** $-1+i\sqrt{3}, -1-i\sqrt{3}$ **13.** $5+5i, 5-5i$ **14.** $-3+8i, -3-8i$
15. $\frac{1}{2}+7i, \frac{1}{2}-7i$ **16.** $-\frac{2}{3}+i, -\frac{2}{3}-i$

Objective 11.5.1

1. 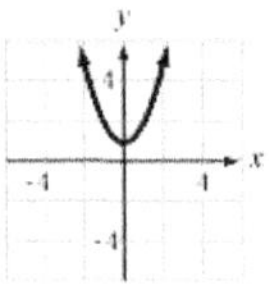
2. 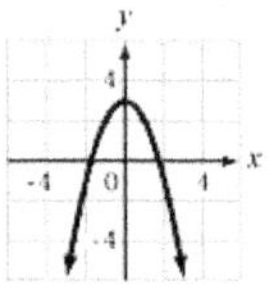
3. 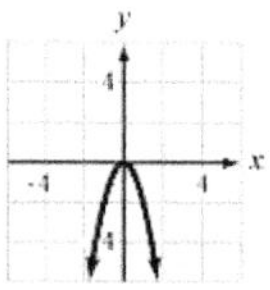
4. 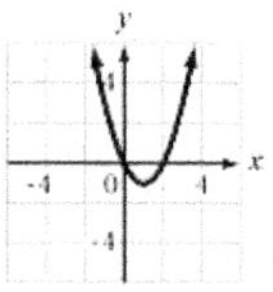
5. 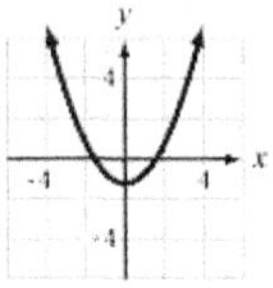
6. 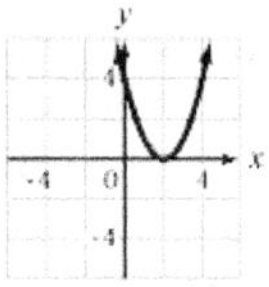
7. 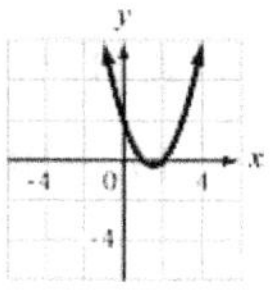
8. 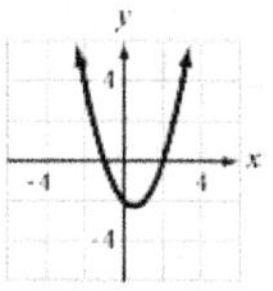

Objective 11.6.1

1. 12 ft, 4 ft **2.** 12 m, 6 m **3.** 4, 6 **4.** 7, 9 **5.** –1, 1, 3 **6.** 4 yrs, 7 yrs
7. 10 ft and 4 ft **8.** 5 m, 8 m **9.** 9 h **10.** smaller: 24 min; larger: 8 min
11. older: 34 min; newer: 30 min **12.** 1st car: 48 mph; 2nd car: 60 mph

Made in the USA
Monee, IL
08 November 2026

68055347R00083